Matrix Metalloproteinases in Health and Disease

Sculpting the Human Body

Matrix Metalloproteinases in Health and Disease

Sculpting the Human Body

Constance E Brinckerhoff, PhD
Geisel School of Medicine at Dartmouth, USA

World Scientific

NEW JERSEY • LONDON • SINGAPORE • BEIJING • SHANGHAI • HONG KONG • TAIPEI • CHENNAI • TOKYO

Published by

World Scientific Publishing Co. Pte. Ltd.
5 Toh Tuck Link, Singapore 596224
USA office: 27 Warren Street, Suite 401-402, Hackensack, NJ 07601
UK office: 57 Shelton Street, Covent Garden, London WC2H 9HE

Library of Congress Cataloging-in-Publication Data
Names: Brinckerhoff, Constance E.
Title: Matrix metalloproteinases in health and disease : sculpting the human body /
Constance E. Brinckerhoff, Geisel School of Medicine at Dartmouth, USA.
Description: New Jersey : World Scientific, 2017. |
Includes bibliographical references and index.
Identifiers: LCCN 2016052213 | ISBN 9789813207547 (hardcover : alk. paper)
Subjects: LCSH: Metalloproteinases. | Proteinase. | Biochemistry.
Classification: LCC QP601.7 .B75 2017 | DDC 572/.76--dc23
LC record available at https://lccn.loc.gov/2016052213

British Library Cataloguing-in-Publication Data
A catalogue record for this book is available from the British Library.

Typeset by Stallion Press
Email: enquiries@stallionpress.com

Printed in Singapore

In memoriam:

Jerome Gross
Edward D. Harris, Jr.
John J. Jeffrey
Stephen M. Krane

For their insight and inspiration.

Acknowledgments

I am forever grateful to Christopher Davis at World Scientific Publishing for contacting me, seemingly out of the blue, and inviting me to oversee this volume on matrix metalloproteinases. He has been unwavering in his support and encouragement, and his enthusiasm for this project is contagious. I am also indebted to Ms. Catherine Yeo Man Ling for her careful editing of all chapters. She is based in Singapore, and so our correspondence has a 12-hour lag. Nonetheless, she always promptly responded to my inquiries with diligence, patience, and excellent advice. I am also deeply appreciative of the chapters contributed by my clinical colleagues, Drs. Dayer, Cao and Libby and their associates. By applying the basic biology of matrix metalloproteinases to a clinical setting, their efforts, along with those of their collaborators have added an important dimension to the book.

Preface

I entered the (not-yet-named) arena of matrix metalloproteinases (MMPs) in January of 1977, when I accepted a position as Instructor in Medicine in the laboratory of Dr. Edward (Ted) D. Harris, Jr., MD at Dartmouth Medical School. I had just completed a 4-year post-doctoral fellowship in Tumor Immunology, also at Dartmouth, and the field of collagenase was a complete mystery to me. Just prior to actually joining the lab, I attended a meeting where Ted hosted a visitor from the Strangeways Laboratory in Cambridge, England. The discussions were all about the biochemistry of collagenase. I was dumbfounded, understanding next to nothing.

At the time, the field was essentially still in its infancy. It was barely 15 years since Jerome Gross (Figure 0.1) had published his landmark paper on tadpole collagenase, and there were still only a handful of investigators in the field. Ted's lab was small, with a nucleus of only three or four people. Slowly, I learned. I watched Carol Vater run gel filtration and DEAE columns, diligently purifying "activator" of latent collagenase, which would ultimately be renamed "stromelysin" or "MMP-3". David Woolley joined us one summer — as he used elegant immune fluorescent technology to demonstrate collagenase in human synovial fibroblasts cultured from tissues excised from patients. I busied myself making multinucleated giant cells from rabbit synovial fibroblasts by treating the cells with polyethylene glycol (PEG), which somehow turned on the collagenase gene and mimicked the behavior of the giant cells found in rheumatoid synovial tissues. Soon, Hideaki Nagase arrived in Hanover to lead work on the synthesis and secretion of collagenase.

Figure 0.1. Jerome Gross, MD, 1917–2014.

In those days, before the advent of today's powerful targeted antirheumatologic therapies, synovectomies performed on patients with rheumatoid arthritis contained copious amounts of proliferative synovial tissue that could be cultured *in vitro*, and then used for enzyme purification or for studies on cell biology. We prepared guinea pig collagen as the substrate for our studies with human collagenase — "bucket chemistry" at its very best. Acid extractions and salt precipitations, with everything kept at 4°C all the time! Warming could denature the collagen, making it unsuitable for use in our collagenase assays. (Today, one simply orders purified collagen already prepared). Under Ted's watchful and critical eye, we were a productive and close-knit group.

On one occasion, I met Jerry Gross, and he kindly invited me to Boston to give a seminar on my own work, which was still very much in its developmental stages. But Jerry was encouraging and kind, and I have never forgotten that visit. Then, in the early 1980s, Ted and Hideaki moved to the University of Medicine and Dentistry of New Jersey, and Carol entered a PhD program. In 1983, I set up my own lab at Dartmouth.

Even then, 20 years after Jerome Gross' initial publication, interest in "vertebrate collagenases" took a while to build. PubMed shows that in 1964 there were a mere 6 articles listed under this topic, 16 in 1965, and 24 in 1967. Exact numbers of citations are difficult to come by due to overlap in key words and misnomers with other enzymes that may have had some collagenolytic activity (e.g., cathepsins). The decade between 1970 and 1980 shows 1950 citations, and this climbs to over 4000 in the following 10 years, and to more than 11,600 from 1990 to 1999. Meanwhile, other enzymes (MMPs) that degraded other components of the extracellular matrix were discovered, fueling research in this area. The search terms for "gelatinase" and "stromelysin" first appeared in 1987, with two listings, and "MMPs" was first found in 1980. How many enzymes were there? What were their substrates? Was their redundancy in their functions? What were their functions? Did they have roles in normal physiology or were their activities confined to pathological conditions? How were they regulated?

Interest in, and answers to, some of these questions occurred in concert with the growth of science and technology in many other areas of study. Over the past 35+ years, I have been privileged to witness (and hopefully contribute to) new knowledge in this important arena. Advances in protein biochemistry, crystallography, molecular biology and gene expression, cell biology, and signal transduction have all been visited upon MMPs, just as they have been on other proteins and subjects of biology.

By now, we have learned a good deal about the basic biology and biochemistry of MMPs. We know that they belong to a larger family called the Metzincins, of which ADAMS and ADAMTS are also members. However, this volume will be confined strictly to MMPs. In the fall of 2016, a PubMed search of the topic MMPs yields more than 46,000 articles, of which more than 4000 are reviews. Clearly, these enzymes are subjects of major interest to many investigators in many areas of study, as scientists and physicians try to apply knowledge and information about MMPs to their own areas of study. Thus, the goal is to summarize the salient features and functions of the MMP family members and then, importantly, to apply this information in a practical manner to understanding how MMPs contribute to normal physiology and to pathology of selected diseases. This volume is not directed at "MMP-ologists", but rather, to those investigators who find themselves needing and/or wanting to apply aspects of MMP biology and biochemistry to their work and thus, need some basic knowledge in order to do this cogently.

It is also meant for clinicians who seek to understand how MMPs contribute to disease pathology and whether and how MMPs are drug targets, or antitargets. Accordingly, the Chapters by noted clinicians (Jean-Michel Dayer, MD in rheumatology; Jian Cao, MD in oncology; and Peter Libby MD in cardiology) represent important practical and clinically oriented contributions. The biochemistry and cell biology of MMPs are not necessarily straightforward, but basic information on the history of these enzymes, their myriad of functions that extend far beyond cleaving of the extracellular matrix, and the complex mechanisms that control their expression will hopefully be valuable to basic scientists and clinician scientists, alike.

Constance E. Brinckerhoff, PhD
Geisel School of Medicine at Dartmouth
October 2016

Contents

Acknowledgments vii

Preface ix

Chapter 1 A Brief Introduction to Proteinases 1
Constance E. Brinckerhoff

Chapter 2 Historical Perspective 13
Constance E. Brinckerhoff

Chapter 3 Regulation of Matrix Metalloproteinase (MMP) Gene Expression 31
Constance E. Brinckerhoff

Chapter 4 "Matrix Metalloproteinases: They're Not Just for Matrix Anymore!" 65
Constance E. Brinckerhoff

Chapter 5 Matrix Metalloproteinases (MMPs) in Homeostasis and Health: Can MMPs be Good for You? 95
Constance E. Brinckerhoff

Chapter 6 Matrix Metalloproteinases (MMPs) and Cytokines in Rheumatology 123
Jean-Michel Dayer, Samantha J. Williamson, Adam P. Croft, Christopher D. Buckley and Carlo Chizzolini

Chapter 7	Matrix Metalloproteinases (MMPs) as Cancer Therapeutic Targets *Jillian Cathcart, Vincent M. Alford and Jian Cao*	157
Chapter 8	Matrix Metalloproteinases in Cardiovascular Diseases *Merry L. Lindsey and Peter Libby*	187
Chapter 9	Matrix Metalloproteinases (MMPs) as Therapeutic Targets *Constance E. Brinckerhoff*	227
Chapter 10	Through the Looking Glass into the Era of Precision Medicine *Constance E. Brinckerhoff*	253
Glossary		267
Index		277

1. A Brief Introduction to Proteinases

Constance E. Brinckerhoff

Exactly how proteinases act on their various protein substrates is a critical component in almost all aspects of normal physiology and disease pathology.[1] Therefore, any discussion on matrix metalloproteinases (MMPs) must place these enzymes in the context of other classes of proteinases in order to fully understand and appreciate their contributions to biology. Many proteinases show exquisite specificity in targeting their protein substrate, whereas others are far more promiscuous, showing broad degradative activity against many substrates.[1]

Data assembled from a variety of public and private sequencing databases and from the MEROPS, InterPro, and Ensemblrecords have identified at least 550 genes that encode proteases in the human genome, often with murine counterparts.[1] Of these, nearly 100 are enzymatically inactive due to substitutions in amino-acid residues that are in regions of the proteinase that are needed for the enzyme to retain proteolytic activity. Interestingly, it has been suggested that these inactive homologues have important regulatory or inhibitory molecules, by functioning as dominant negatives or by "sopping up" inhibitors from the local environment, and thus permitting increases in proteolytic activity.[1] This intriguing concept implies additional and novel functions for proteases, beyond their direct enzymatic actions, a concept that has been realized with MMPs.

Five classes of proteinases have been designated based on mechanisms of catalysis: aspartic, cysteine, threonine, serine, and metalloproteinases.[1,2] The metalloproteases and serine proteinases have the most members, with 186 and 176 enzymes, respectively, followed by the cysteine proteases, of which there are 143 enzymes (Figs. 1.1 and 1.2).[3] The threonine and aspartic proteases have very specific functions and are less common and have only 27 and 21 members, respectively.[1] The aspartic and metalloproteinases use an activated water molecule to mediate a nucleophilic attack[1] on the peptide bond of the substrate, while in the cysteine, serine, and threonine proteinases, the nucleophile is an amino acid (cysteine, serine, or threonine) located in the active site of the enzyme. For the degradation of matrix proteins, these proteinases are primarily "endopeptidases", that is, they cleave peptide bonds within the matrix molecule. They are synthesized as inactive precursors, or "pro-enzymes", which must be activated proteolytically either by autocrine or paracrine mechanisms. The proform of endopeptidases is a critically important regulatory step that prevents indiscriminate and rampant proteolytic activity.[1,4]

1.1 Aspartic Proteinases

These proteinases have two aspartic acid residues as an integral component of their catalytic site.[1,4] Aspartic proteinases are active at acidic pH and inactive at neutral pH. Mostly, they are enzymatically active extracellularly in the pericellular space, where the pH is sufficiently low (pH 3.5–5.5). Intracellularly, they are active in lysosomes or in a microenvironment where the pH is between 5.5 and 7.0. Examples of aspartic proteinases include pepsin, renin, HIV protease, and the lysosomal cathepsins D and F. Aspartic proteinases have broad substrate specificities. However, they function within in a somewhat narrow biological context, which is defined by optimal pH and their own particular ability to target a specific substrate.

1.2 Cysteine Proteinases

Cysteine proteinases also function optimally at an acid pH and, like the aspartic proteinases, they are usually stored in lysosomes.[1,4] These proteases have a catalytic mechanism that involves a nucleophilic

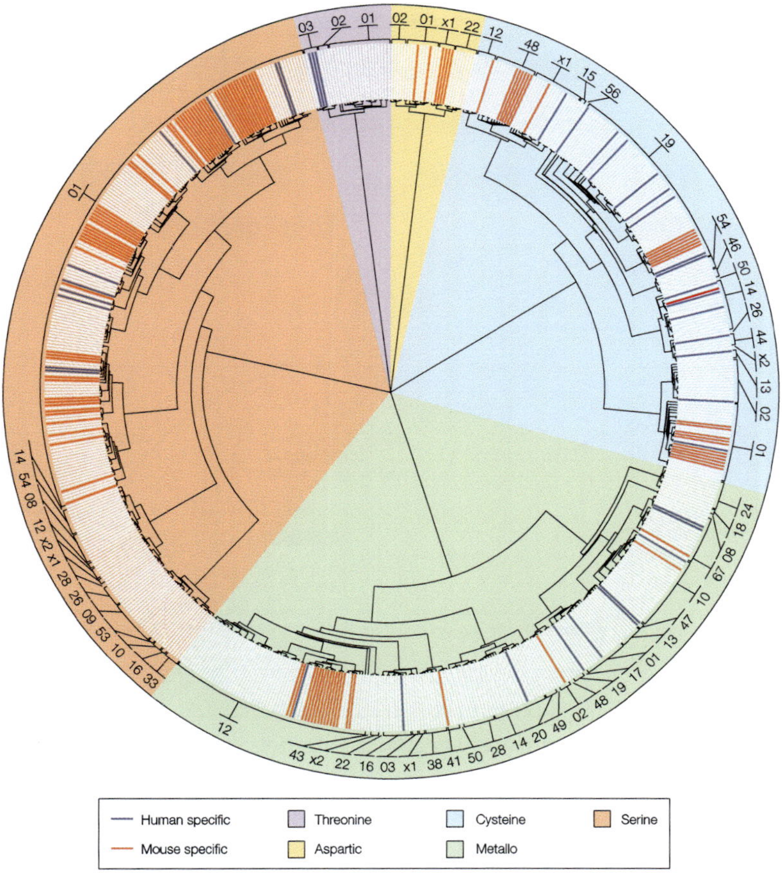

Figure 1.1. The protease wheel. Unrooted phylogenetic tree of human and mouse proteases. Proteases are distributed in 5 catalytic classes and 63 different families. The code number for each protease family is indicated in the outer ring. Protein sequences that correspond to the protease domain from each family were aligned using the ClustalX program. Phylogenetic trees were constructed for each family using the Protpars program. A global tree was generated using the protease domain from one member of each family, and individual family trees were added at the corresponding positions. The figure shows the nonredundant set of proteases. Orthologous proteases are shown in light gray, mouse-specific proteases are shown in red, and human-specific proteases in blue. Metalloproteases are the most abundant class of enzymes in both organisms, but most lineage-specific differences are in the serine protease class, making this sector wider. The 01 family of serine proteases can be divided into 22 smaller subgroups on the basis of involvement in different physiological processes, to facilitate the interpretation of differences. (From Puente *et al.* 2003.)

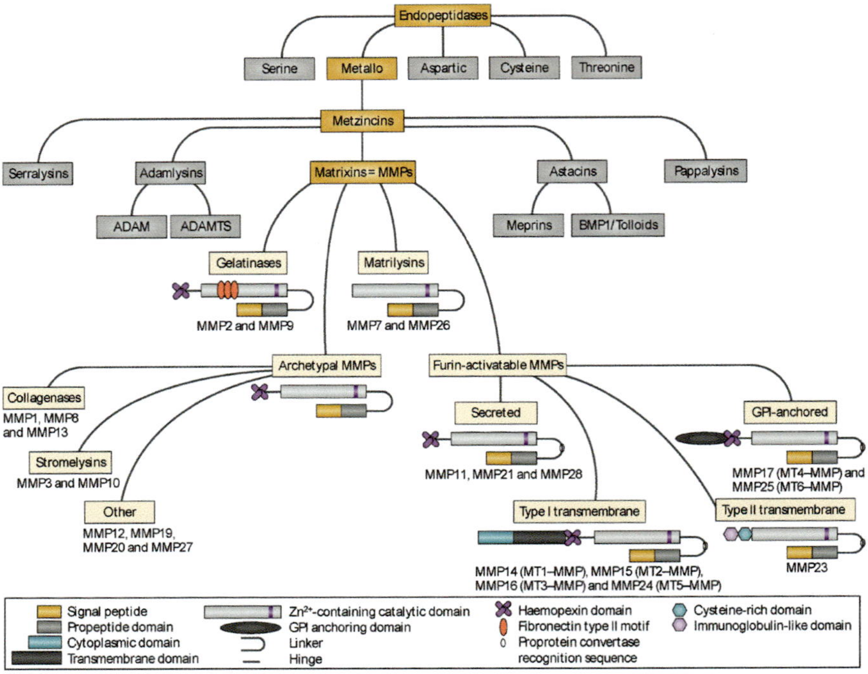

Figure 1.2. Schematic overview of the structure of MMP family members and the relationship with other metzincin superfamily members. MMPs share a common domain structure: the predomain that Nature Reviews | Drug Discovery contains a signal peptide responsible for secretion; the pro-domain that keeps the enzyme inactive by an interaction between a cysteine residue and the Zn^{2+} ion group from the catalytic domain; and the hemopexin-like carboxy-terminal domain, which is linked to the catalytic domain by a flexible hinge region. MMP7 and MMP26 lack the hinge region and the hemopexin domain. MMP2 and MMP9 contain a fibronectin type II motif inserted into the catalytic site, and MT-MMPs have a transmembrane domain or a glycosylphosphatidylinositol (GPI) anchor at the C terminus. MMP23 has unique features: the amino-terminal signal anchor that targets MMP23 to the cell membrane, a cysteine array, and an immunoglobulin-like domain. ADAM, a disintegrin and metalloproteinase; ADAMTS, ADAM with thrombospondin motifs; BMP1, bone morphogenetic protein 1. (From Vandenbroucke and Libert 2014.)

cysteine thiol in a dyad or triad, that is, two or three amino-acid residues that function together at the center of the active site. Cysteine proteases have several roles in extracellular matrix (ECM) remodeling. Lysosomal cysteine proteinases secreted from osteoclasts facilitate bone resorption by degrading collagen. They are often involved in the digestion of phagocytosed materials, giving them a prominent role in activated macrophages. Cathepsins B, L, N, and S and the calpains are a few examples of cysteine proteinases. Other cysteine proteinases are Cathepsins B and L, which may be present extracellularly, especially in activated macrophages. These cathepsins can cleave the nonhelical regions of fibrillar collagens at acid pH, leading to depolymerization of collagen fibrils. Hence, they are sometimes referred to as "collagenases" although they do not meet the classical definition, which is "an enzyme that catalyzes the hydrolysis of peptide bonds in triple helical regions of collagen".[5] Cathepsins L and S also attack elastin, giving them a role in ECM metabolism. Thus, cysteine proteinases are ubiquitous and can contribute substantially to matrix remodeling in health and disease, either alone or in combination with other classes of proteinases.

1.3 Threonine Proteinases

Threonine proteinases are a more recently discovered category of proteinases and they function as a component of proteasomes. Proteasomes are the main mediators of intracellular degradation of a wide variety of cellular proteins and they are implicated in several physiological and pathological cellular functions.[6,7] The proteasome pathway is involved in matrix degradation by several mechanisms, which include both transcriptional and posttranslational mechanisms. By controlling the concentration and turnover of several ECM components, the proteasome pathway contributes to extracellular proteolytic events by modulating the expression and activity of MMPs and their endogenous inhibitors (see the following). Therefore, since matrix remodeling and degradation can be controlled by proteasome activities, proteasome modulation might be a novel therapeutic strategy in some pathologic conditions[7] and perhaps future studies may reveal additional mechanisms of their action and therapeutic strategies that target them.

1.4 Serine Proteinases

Serine proteinases comprise a large family of enzymes, with a broad range of substrates. Consequently, they participate in many important biological processes such as digestion, blood clotting, and immune response.[1,4] In contrast to aspartic and cysteine proteinases, they function at neutral pH, cleaving peptide bonds in proteins where serine serves as the nucleophilic amino acid at the active site of the enzyme. Examples of serine proteinases are enzymes involved in digestion (trypsinogen/trypsin, chymotrypsinogen/chymotrypsin), blood clotting (plasminogen/plasmin, prothrombin/thrombin), and homeostasis (tissue and plasma prekallikrein/kallikrein, and tissue remodeling (proelastase/elastase). Further, many serine proteinases are important by acting indirectly on the ECM by activating latent MMPs, cleaving them to their enzymatically active form. Some serine proteinases are stored in azurophil granules of polymorphonuclear (PMN) leukocytes, while others (i.e., tryptase, chymase, and cathepsin G) are secreted. Plasminogen activators (PAs), tissue PA and urokinase PA, are secreted by several cell types, including fibroblasts, chondrocytes, and tumor cells, and convert plasminogen to plasmin. Plasmin then degrades fibrin as well as ECM components such as aggrecan, type IV collagen, and laminin. Kallikrein binds to receptors on fibroblasts, macrophages, and tumor cells, where it helps to coordinate several physiological functions, including blood pressure, liquification of semen liquefaction, and shedding of skin. Serine proteinases are abundant, with many physiologic functions. Given their ability to cleave many substrates and to activate latent MMPs, serine proteinases are important adjuncts to MMPs in ECM degradation.

1.5 Matrix Metalloproteinases

MMPs, sometimes called matrixins, belong to the super family of metzincins,[2,4,8–11] which is characterized by a catalytic zinc atom at the active site of the enzyme, followed by a conserved methionine (Fig. 1.3). Metzincins contain several subfamilies, including two that have activities related to those of the MMPs, that is, a disintegrin and metalloproteinase (ADAMS) and a disintegrin and metalloproteinase with thrombospondin motif (ADAMTS), metalloproteinases that were first reported in 1992 and 1997, respectively.[2,12] ADAMs are membrane-associated enzymes, which are expressed by

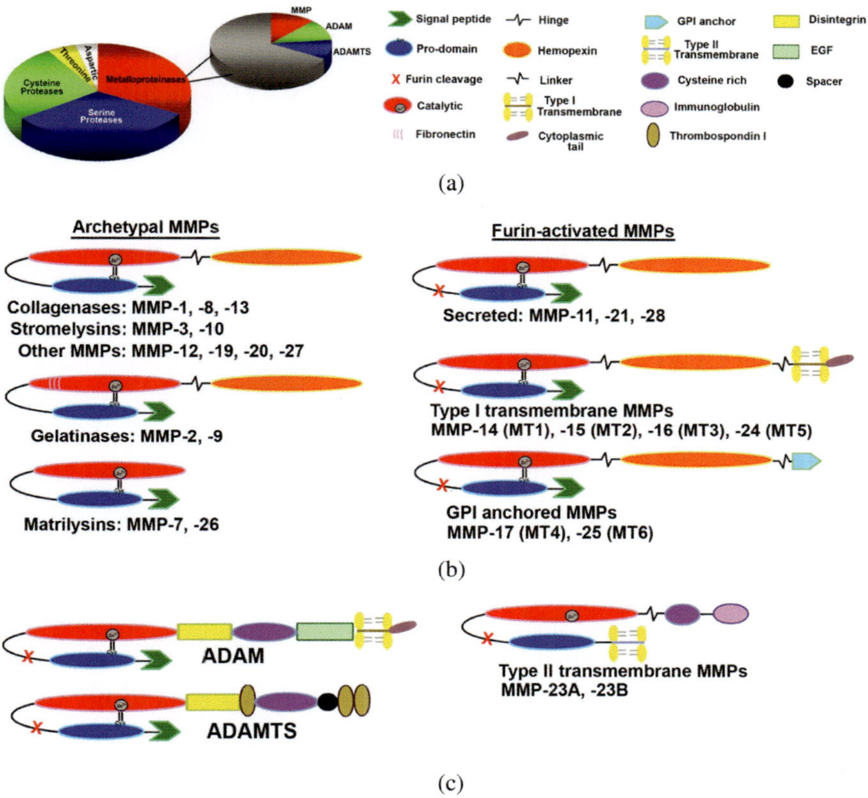

Figure 1.3. Protease classification and structure. (a) The human "degradome", the repertoire of proteases produced by cells, consists of at least 569 proteases and homologues subdivided into five classes: 21 aspartic, 28 threonine, 150 cysteine, 176 serine, and 194 metallo-proteases, including MMP, ADAM, and ADAMTS family members. (b) Structural classification of MMPs based on domain composition, including secreted and membrane-associated MMPs, and MMPs that are activated intracellularly via furin-mediated cleavage. (c) Most ADAMs are type I transmembrane protein that possess disintegrin, cysteine rich, and EGF domains in lieu of the MMP hemopexin domain. ADAMTSs are secreted proteins that contain thrombospondin I motifs in lieu of the EGF domain. (From Rivera *et al.* 2010.)

a wide variety of cell types, and are involved in functions as diverse as sperm–egg binding, myotube formation, neurogenesis, and proteolytic processing of cell surface proteins, giving them the nickname "sheddase".[2,12] A particularly well-known sheddase is tumor necrosis factor-α (TNF-α) converting enzyme (TACE), a unique member of the ADAM family that cleaves this membrane-bound enzyme. The enzymatic activities of ADAMs are varied and include (1) cleaving collagen propeptides, (2) inhibiting angiogenesis, (3) degrading cartilage proteoglycans, and (4) maintaining blood coagulation homoeostasis as the proteinase that cleaves von Willebrand factor. Both ADAMS and ADAMSTS are increasing subjects of study, and many of their functions and activities either overlap and/or complement those of MMPs.

However, it is the MMPs that have been most studied. The MMPs are a large family of zinc-dependent endopeptidases that, collectively, can degrade all components of the extracellular matrix. Although MMPs were originally defined by their ability to degrade the ECM, it is now very clear that these enzymes have a myriad of additional and important functions in biology (see Chapter 4).[1–4,8–11,13,14] Along with their dependency on zinc for enzymatic activity, MMPs also require Ca^{++}. While most MMPs are secreted, several MMPs are membrane-type MMPs (MT-MMPs). All MMPs share a common three-domain structure: the pro-peptide, the catalytic domain, and the hemopexin-like C-terminal domain, which is linked to the catalytic domain by a flexible hinge region (Fig. 1.4).[13] Like the other classes of endopeptidases, MMPs are synthesized as inactive zymogens, and in most instances, the pro-peptide domain must be cleaved for enzymatic activity. This pro-peptide is part of the "cysteine switch",[3,4,9,11,13,14] a conserved cysteine residue that interacts with the zinc in the active site and by binding to it, keeps the enzyme in an inactive form. In most MMPs, the cysteine residue lies within the conserved sequence PRCGxPD. When the enzyme is activated, the interaction between cysteine-Zn++ is broken, and this allows a water molecule to bind to the zinc and facilitate peptide hydrolysis. In contrast, however, the MT-MMPs contain a cleavage site for the furin-like pro-hormone convertase as part of this domain. When cleaved, it activates the enzyme intracellularly, so that the MT-MMPs are already active when they are inserted into the membrane.

Whether they are secreted or membrane-bound, MMPs are the major class of endopeptidases that degrades the ECM. They were

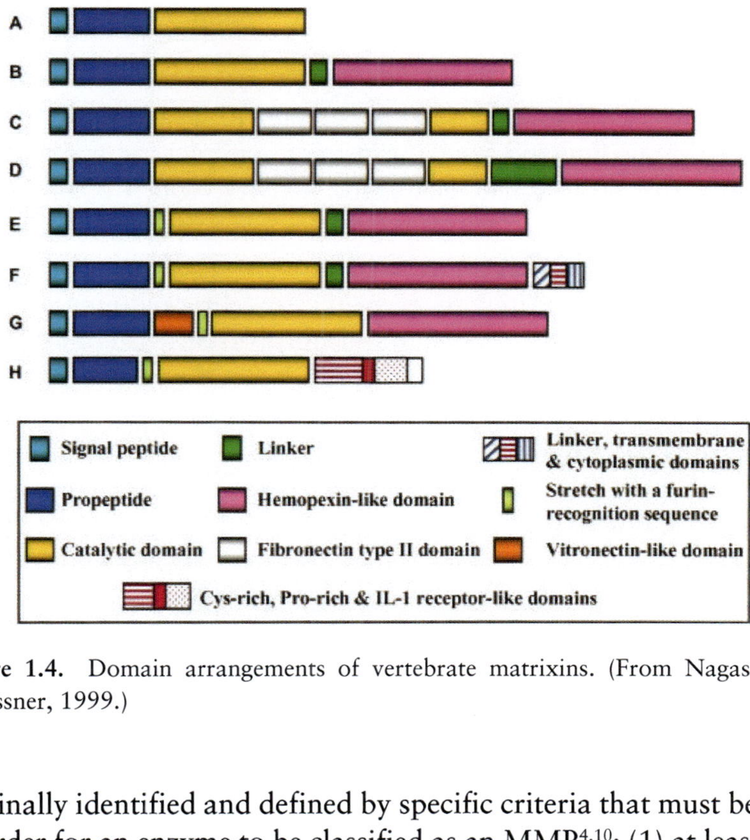

Figure 1.4. Domain arrangements of vertebrate matrixins. (From Nagase and Woessner, 1999.)

originally identified and defined by specific criteria that must be met in order for an enzyme to be classified as an MMP[4,10]: (1) at least one type of ECM molecule is proteolytically degraded; (2) catalysis depends on zinc at the active site; (3) zymogen is activated by proteinases or by organomercurials; (4) the active enzyme is inhibited by ethylenediaminetetraacetic acid (EDTA), 1,10 phenanthroline, and by one of the tissue inhibitors of metalloproteinases (TIMPs); and (5) the cDNA has sequence homology to collagenase (MMP-1).[4] Although an MMP must be capable of degrading a component of the ECM, it is now well accepted that MMPs have many other functions. Consequently, they are implicated as major players in many aspects of biology and pathology. Nonetheless, it is essential to remember that MMPs are often dependent on interactions with other classes of endopeptidases to optimally mediate their effects.

1.6 Endogenous Inhibitors of Endopeptidases

Given the enormous diversity of endopeptidases (Fig. 1.1) and their huge variety of substrates, it is not difficult to understand the need for endogenous biologic mechanisms to counteract the potential for uncontrolled and rampant enzymatic activity, which could wreak havoc in biology. The most prominent endogenous inhibitor is nonspecific: the very large plasma glycoprotein (725kD), alpha$_2$ macroglobulin (α_2M).[4,9] Alpha$_2$M is synthesized by predominantly the liver, also by local macrophages and fibroblasts. It is comprised of four identical subunits linked by disulfide bonds, and it effectively and irreversibly inactivates all endoproteinases. Alpha$_2$M uses a region of its molecule called the "bait" to attack proteinases; proteolytic cleavage of this "bait" region causes a structural change in the inhibitor, trapping the proteinase within the α_2M molecule. Alpha$_2$M will not bind inactive enzymes. This nonspecific proteinase inhibitor is an important first line of defense against pathologic matrix degradation, but its large size precludes it from perfusing into tissues. Thus, its role *in vivo* is largely confined to the bloodstream. Nonetheless, its potency and ubiquitous presence in plasma/serum send a cautionary note when investigators plan *in vitro* experiments to monitor active proteinases in culture medium with serum.

In addition to the nonspecific α_2M, there are also specific inhibitors of endopeptidases.[1,2,4,9] Active serine proteinases are blocked by serpins and by aproprotinin. Cysteine proteinases are inhibited by cystatins, and nearly all aspartic proteases are inhibited by pepstatins. The activities of MMPs are specifically blocked by TIMPs, of which there are four family members, TIMP-1, –2, –3, –4. Their molecular weights range from 22 to 29kDa, with TIMPs 1 and 3 containing carbohydrate, while TIMPs 2 and 4 do not. The connective tissue cells that make MMPs, such as fibroblasts, also produce TIMPs. TIMPs inhibit MMP activity by noncovalent binding to the active site of the target MMP with 1:1 stoichiometry. A conserved cysteine residue at position 1 of the TIMP chelates the MMP active site zinc ion and expels the essential water molecule. The tertiary structure of all TIMPs is conserved, and therefore, all TIMPs can inhibit all MMPs, although TIMP-1 is a poor inhibitor of MT1-MMP.

Since TIMPs can block the activity of MMPs, and are produced endogenously, the concept of using them as therapeutic agents is appealing. This observation has led to the suggestion of targeted overexpression of TIMPs as a clinical therapy in diseases with overexpression of MMPs, such as rheumatoid and osteoarthritis.[9,14] However, in addition to blocking MMP enzyme activity, TIMPs have other effects. TIMP-2 can associate with latent proMMP-2 and is required for the activation of the proteolytic activity of MMP-2. TIMP-1 associates with the carboxy-terminal domain of progelatinase B (MMP-9), blocking its activation by stromelysin-1. In addition, overexpression of TIMP can have paradoxical effects, such as increased cell growth and invasion, decreased angiogenesis, and increased apoptosis, raising serious questions about target specificity.

1.7 Summary and Conclusion

These five classes of proteinases and their inhibitors do not function in a vacuum or in an isolated environment. Several or all classes of proteinases and their inhibitors are often present in varying concentrations in many normal tissues, as well as in diseases. In the end, it is the complex interactions among this pot-pourri of biologically active molecules within a tissue microenvironment and within the whole organism that crucially influence our biology and our pathology.

References

1. Puente, XS, LM Sánchez, CM Overall and C López-Otín (2003). Human and mouse proteases: a comparative genomic approach. *Nature Reviews Genetics*, 4, 544–558.
2. Rivera, S, M Khrestchatisky, L Kaczmarek, GA Rosenberg, and DM Jaworski (2010). Metzincin proteases and their inhibitors, foes or friends in nervous system physiology? *Journal of Neuroscience*, 30, 15337–15357.

3. Vandenbroucke, RE and C Libert (2014). Is there new hope for therapeutic matrix metalloproteinase inhibition? *Nature Reviews Drug Discovery*, 13, 904–927.
4. Nagase, H and Y Okada (1997). Proteinases and matrix degradation. In *Textbook of Rheumatology,* WN, Kelley, S Ruddy, ED Harris and CB Sledge (eds.), 5th Ed., pp. 323–341. Philadelphia: WB Saunders.
5. *Dorland's Illustrated Medical Dictionary*, 32nd Edition, (2011). Philadelphia: Saunders.
6. Baumeister, W, J Walz and E Seemu (1998). The proteasome: paradigm of a self-compartmentalizing protease. *Cell,* 92, 367–380.
7. Syggelos, SA, AJ Aletras, I Smirlaki, and SS Skandalis (2013). Extracellular matrix degradation and tissue remodeling in periprosthetic loosening and osteolysis: focus on matrix metalloproteinases, their endogenous tissue inhibitors, and the proteasome. *BioMed Research International,* 2013, 18 p.
8. Gomis-Rüth, FX (2009). Catalytic domain architecture of metzincin metalloproteases. *Journal of Biological Chemistry,* 284, 15353–15357.
9. Murphy, G and H Nagase (2008). Progress in matrix metalloproteinase research. *Molecular Aspects of Medicine*, 29, 290–308.
10. Iyer, RP, NL Patterson, GB Fields and ML Lindsey (2012). The history of matrix metalloproteinases: milestones, myths, and misperceptions. *Amercian Journal of Physiology Heart and Circulatory Physiology,* 303, H919–H930.
11. Fanjul-Fernández, M, AR Folgueras, S Cabrera and C López-Otín (2010). Matrix metalloproteinases: evolution, gene regulation and functional analysis in mouse models. *Biochimica Biophysica Acta,* 1803, 3–19.
12. Edwards, DR, MM Handsley and CJ Pennington (2008). The ADAM metalloproteinases. *Molecular Aspects of Medicine*, 29, 258–289.
13. Nagase, H and JF Woessner (1999). Matrix metalloproteinases. *Journal of Biological Chemistry,* 274, 21491–21494.
14. Brinckerhoff, CE and LM Matrisian (2002). Matrix metalloproteinases: a tail of a frog that became a prince. *Nature Reviews Molecular Cell Biology*, 3, 207–214.

2. Historical Perspective

Constance E. Brinckerhoff

2.1 The Age of Discovery

The most abundant proteins in our bodies are the interstitial collagens, types I, II, and III. Although there is a host of other collagen types, which are present in smaller amounts, the interstitial collagens comprise about 30% of the total protein.[1–3] The three interstitial collagens are the primary components of the extracellular matrix, forming the major structural proteins of skin, bone, cartilage tendon, and blood vessels.[4] All collagens have an iconic structure: a triple helix where the three chains are tightly wound around each other, with the sequence gly-x-y, where gly is present at every third residue and x and y are often proline and hydroxyproline. The small size of glycine results in the three chains being compressed together, which helps to give collagen its hallmark tensile strength. This characteristic triple helical structure has come to define both collagen and the unique nature of the enzyme(s) that cleave it.

Collagen types I and III are the most ubiquitous and make up our visceral connective tissues and bones.[5] Type II collagen is unique to cartilage, and type IV collagen comprises basement membrane and thus has a critical role in the endothelium and blood vessels.[5] Considering the major role of collagen as a dominant structural protein in our bodies, there has been interest in understanding the mechanisms by which collagen molecules were degraded. While it had long been known that collagenases isolated from bacteria (such as *Clostridium histolyticum*) readily cleave the collagen molecule in numerous places, the existence of a homologue in eukaryotic/mammalian tissues remained a mystery worthy of extensive investigation and speculation.

How, for example, did the postpartum uterus so quickly return to its normal size and weight? Measurements of collagen content revealed a rapid loss but how this was mediated was not clear. In 1962, Frederick Woessner described proteolytic activities in the complex experimental system of the involuting rat uterus, studies which were suggestive of enzymes that optimally functioned at pH 5.5 and 8.5.[6] Cathepsin-like activity was mentioned, but other possibilities, such as "peptidases able to digest unusual peptides...that might be expected to arise from collagen" were suggested. Dr. Woessner was clearly on the right track, and perhaps the "activity" he saw at pH 8.5 reflected remnants of what would eventually become matrix metalloproteinases (MMPs).

Later that same year, Gross and Lapiere published the definitive paper describing a "collagenolytic activity in amphibian tissues" in a tissue culture system.[7] "Cultivating bullfrog tadpole tissues (skin, gut and gills) on thermally reconstituted neutral calf and guinea pig skin collagen gels resulted in degradation of the collagen substrate into dialyzable collagen peptides. The collagen substrate was not degraded by other common proteolytic enzymes, such as cathepsin C, or by extracts of tadpole tissues" (Fig. 2.1). These findings suggested that the collagenolytic "activity" was secreted extracellularly and that it might represent a newly discovered enzyme.

As the early stages of investigation progressed, Gross and colleagues remarked that "the isolation and characterization of this enzyme system...[proved] most elusive".[8] Nagai, Lapiere, and Gross were finally successful, being the first to purify the tadpole enzyme,[9] perhaps because the experimental system was comparatively simpler than that of mammalian tissues, such as the rat uterus of Dr. Woessner. Even so, Gross and colleagues found purification a challenge, always needing to rule out the presence of contaminating bacterial collagenase: "contaminating cultures always reveal large numbers of bacteria upon microscopic examination and were discarded".[9] Further, "extracts of tissue...failed to yield an enzyme which, at neutral pH and physiological temperature, will attack

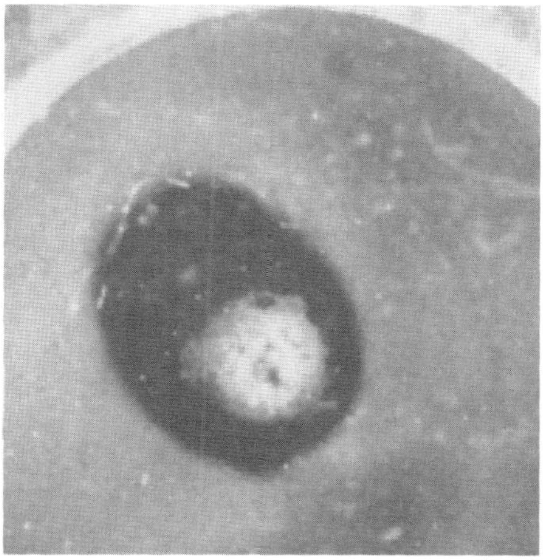

Figure 2.1. Degradation of collagen by the tadpole tail. Placing a slice of tadpole tail (center) on a collagen gel results in a zone of clearing that surrounds the tissue. This indicates collagenase activity. The image was produced by John Jeffrey and kindly provided by George Stricklin, Vanderbilt University, USA. (From Brinckerhoff and Matrisian, 2002.)

native collagen". Their conclusion: the "activity" was secreted into the surrounding environment and allowed the diffusion of a collagenolytic principle away from the tissue.[7] They learned early on, but probably did not understand, why bovine serum had a marked inhibitory effect[9]; the alpha$_2$ macroglobulin (α_2M) in serum is a potent inhibitor of all proteases. Thus, serum-free medium was essential, a tenant that is still vitally important today.

Eventually, they assigned the following characteristics to their "activity": (1) it was reversibly inhibited by ethylenediaminetetraacetic acid (EDTA) and irreversibly by cysteine; (2) it was active at neutral pH; and (3) compared to bacterial *C. histolyticum*, there were radical differences in its mode of attack on the collagen molecule, with only a single cleavage in the triple helix. Several studies had reported that at temperature below 30°C, the collagen molecule is severed to produce two fragments, the 3/4 TCA and the 1/4 TCB (Fig. 2.2).[9] Remarkably, only three peptide bonds were cleaved from the 300,000-MW species, suggesting one in each of the three collagen

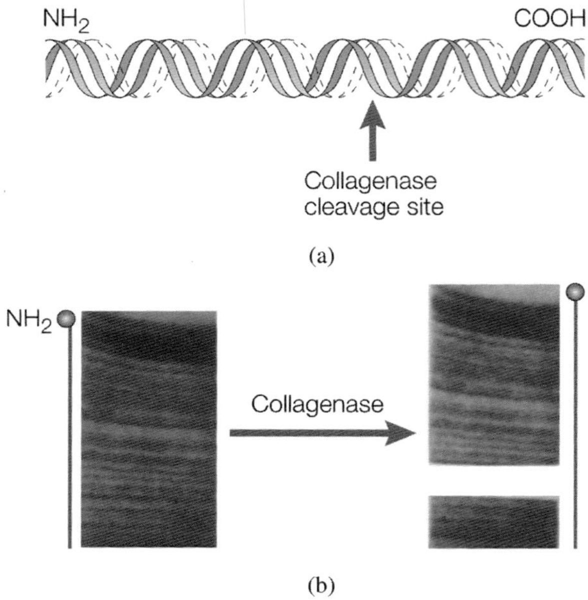

Figure 2.2. Degradation of interstitial collagen by collagenase. (a) A monomer of types I, II, or III interstitial collagen, which shows the position of the glycine 775–leucine/isoleucine 776 bond that is cleaved by MMP-1. (b) Cleavage of collagen by frog collagenase into 3/4:1/4-length fragments, as observed by reconstituting these fragments into segment long spacing structures that are viewed with an electron microscope. (From Brinckerhoff and Matrisian, 2002.)

chains.[9] Consequently, Gross and colleagues suggested that "a more inclusive and useful definition of a collagenase might be an enzyme which breaks covalent bonds within the polypeptide backbone of the collagen molecule, causing a change in size, shape configuration of stability under conditions which will not in themselves denature the protein...since it is now likely that the animal collagenases differ considerably in their properties and mode of action and **may constitute a much larger and diverse group**".[9]

Indeed, a handful if investigators were about to embark on studies that would identify a host of related enzymes, with a variety of "activities" against different matrix molecules. But the initial and exciting observation by Gross and Lapiere was followed by years of difficult work as researchers struggled to purify this collagenolytic "activity" from mammalian tissues. These studies were especially challenging since the investigators did not know that the enzyme was synthesized and secreted as an inactive proenzyme, which could be activated by other proteases (and even, sometimes, by itself) during the process of purification.[10] Thus, activation appeared "spontaneous", perhaps because mammalian tissues with high levels of collagenase also contained high levels of other enzymes that could activate latent collagenase, thereby obscuring the naturally occurring latent state of the enzyme. In 1970, Jeffrey and colleagues successfully purified the enzyme from human skin[11] and then from the involuting rat uterus.[8]

As Fred Woessner had suggested several years earlier, the postpartum uterus was an ideal target tissue to be investigated, primarily because collagenase was so abundant. An enzyme was isolated, partially purified, and characterized, with many characteristics in common with the tadpole enzyme.[11] However, the mammalian enzyme(s) showed distinct immunologic reactivity with the tadpole enzyme and some cross reactivity between rat and human skin/synovial tissues, a finding that hinted at some of the complexities that would come to be associated with the MMP family.

Thus, by the early 1970s, a scant 10 years after the initial publication and with a small, but growing constituency of investigators, numerous reports of mammalian homologues of tadpole collagenase began to appear, and new enzymes were being discovered, sometimes simultaneously.[10] Logical choices of study were tissues in which extensive destruction of connective tissues containing collagen (i.e., cartilage and bone in rheumatoid arthritis) (Fig. 2.3). In 1972, collagenase activity was detected in human rheumatoid synovial tissue and eventually purified from these tissues in 1975.[10,12]

Concomitant with the observations of rampant connective tissue destruction in arthritis was the association between matrix destruction and cancer. Degradation of the basement membrane was a well-recognized symbol of malignancy, and the search was on for the enzyme(s) responsible. In 1980, attention was focused on the degradation of type IV collagen found in basement membrane. There

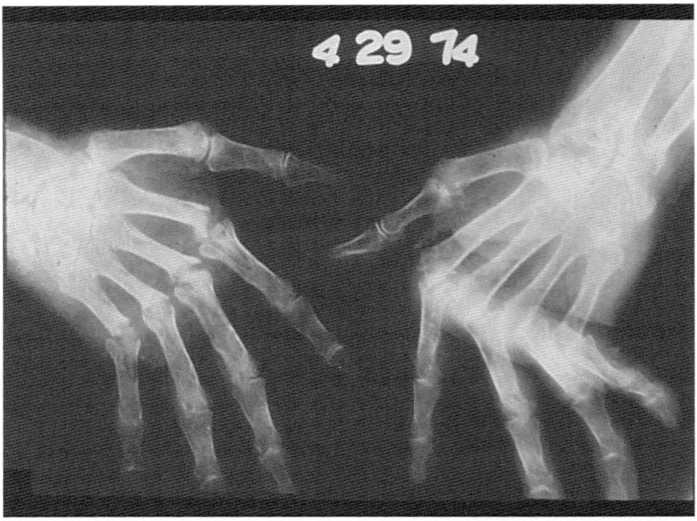

Figure 2.3. Joint destruction in rheumatoid arthritis. (From the collection of Edward D. Harris Jr.)

was speculation on the existence of a specific collagenase, which played an essential role in tumor invasion and metastasis.[13] The "principle" responsible for this degradation was not precisely identified, but when the type IV collagenases/gelatinases MMP-2 and MMP-9 were cloned, the ability to degrade type IV collagen was assigned to them. Then, as more and more *in vivo* studies were carried out in the late 1980s and early 1990s, several other MMPs were shown to contribute to tumor invasion and metastasis, in addition to the gelatinases, which degraded type IV collagen.[10]

Eventually, it became apparent that tadpole collagenase was secreted as a zymogen, requiring proteolytic activation, which reduced its molecular weight (MW) by some 10,000.[14] This was an important discovery, since the prevailing view had been that tadpole collagenase was synthesized *de novo* as an active enzyme.[14] However, careful purification and fractionation studies revealed that activation of the zymogen was accomplished by incubation with collagenase-free medium taken from cultures of tail fin. "The unidentified activator in the medium [was] heat labile and nondialyzable".[14] The documentation of collagenase as a zymogen emphasized the critically important role of latency in regulating enzymatic activity and ignited curiosity about the nature of the activator. The search for the "heat labile and nondialyzable" activating principle was on.

However, in the process of searching for "activator", additional MMP family members were discovered during the late 1970s and early 1980s, with some discoveries duplicated in different laboratories almost simultaneously. This resulted in redundant naming of the same enzyme with "pet" names that reflected similar activities (Table 2.1).[15] For example, what is now MMP-3 was simultaneously named "stromelysin", "proteoglycanase", "procollagenase activating protein", and "transin-1" as it was isolated from several

Table 2.1. Matrix metalloproteinases.

MMP-1	Collagenase, interstitial (fibroblast-type)
MMP-2	Gelatinase, 72-kDa
	Type IV collagenase, 72-kDa
MMP-3	Stromelysin
	Proteoglycanase
	Procollagenase activating protein (CAP)
	Transin-1
MMP-4	Collagen telopeptidase
MMP-5	3/4-Collagenase (= 72-kDa gelatinase), number now vacant
MMP-6	Acid metalloproteinase of cartilage
MMP-7	Small metalloproteinase of uterus
	Pump-1
MMP-8	Neutrophil collagenase (interstitial)
*MMP-9	Gelatinase, 92-kDa
	Type V collagenase
	Type IV collagenase, 92-kDa
*MMP-10	Stromelysin-2
	Transin-2

Source: Nagase *et al.* (1992).

*Newly assigned MMP numbers. Enzyme names without numbers are synonyms for the preceding name.

different labs almost at the same time. Most likely, all of these names reflected the broad substrate specificity that has come to be associated with MMP-3.

Cloning genes was becoming common by the mid-1980s, even though by today's standards, the technology was exceedingly crude. Nonetheless, gene cloning resulted in the rapid isolation and identification of MMPs[10], with the cloning of cDNAs for several MMPs, including rat transin, the human stromelysins, and human and rabbit collagenase, appearing almost simultaneously (Fig. 2.4).[10] New family members were discovered as MMP cDNAs were cloned and sequenced.

As genomic clones corresponding to cDNAs were isolated and characterized, the similarities among sequences began to reveal family relationships.[10] Intriguing thoughts about the possible existence of a gene family began to develop. The genes for many MMP family members were found to be about 10 kb in length, with 10 exons and 11 introns, and were clustered on chromosome 11q22.23, which houses most of the MMPs (–1, –3, –7, –8, –10, –12, –13, –20, and –27).[10] However, the genes for the gelatinases, MMP-2 and MMP-9, were located on chromosomes 16 and 20, respectively, and they contained three additional exons, which encoded fibronectin-like repeats.

Clearly, the family was growing, as were numbers of investigators studying these enzymes. By September 1989, there was more than sufficient interest to support a conference of about 250 attendees at a lovely beach resort in Destin, Florida. The research presented there and the informal discussions ignited additional interest in this seemingly ever-growing family of enzymes that could annihilate the extracellular matrix. It was time to consolidate the burgeoning information into a cohesive and logical terminology: the term MMPs was officially introduced,[15] even though the term had first mentioned several years earlier.[16] The formal definition of MMPs went as follows: "Matrix metalloproteinases are members of a family of zinc metalloproteinases of animal origin that act outside the cell on components of the extracellular matrix and that are homologous to interstitial collagenase".[15]

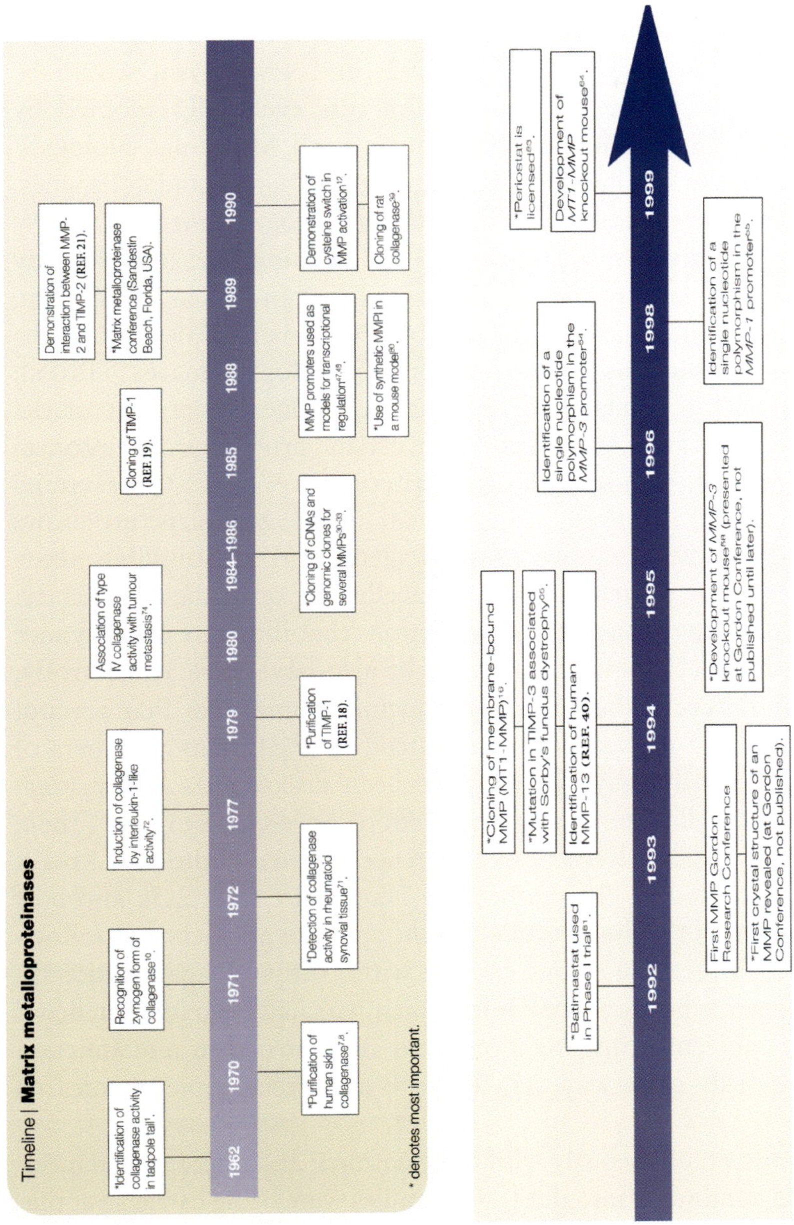

Figure 2.4. MMP timeline.

Source: Brinckerhoff and Matrisian (2002).

As originally suggested and to avoid confusion,[15] each MMP would be assigned a number. Thus, interstitial collagenase would be Matrix Metalloproteinase-1 and would be abbreviated MMP-1. The 72-kDa gelatinase would be MMP-2, and stromelysin would be MMP-3, and so on. This convention was eventually adopted by the International Union of Biochemistry and Molecular Biology.[17] Along with the original criteria delineated by Gross and colleagues, there were several caveats, which reflected progress made in understanding the biochemistry of MMPs: (1) blocking proteinase activity with 1,10-phenanthroline; (2) the proteinase appears in latent form; (3) the latent form can be activated by organomercurials; (4) enzyme activity is inhibited by tissue inhibitor of metalloproteinases (TIMPs); and (5) the proteinase hydrolyzes at least one component of the extracellular matrix, such as collagen, elastin, proteoglycans fibronectin, laminin, or gelatin. Still other criteria were: (1) Ca^{++} is required for activity; (2) zinc is an intrinsic metal ion; (3) gene structure shows homology to the parent enzyme: collagenase; (4) α_2M inhibits activity, and (5) the proteinases act extracellularly or at the cell surface. Finally, there were significant predictions that illustrate clairvoyance and wisdom: (1) these enzymes might also have intracellular roles; (2) their natural substrates might include substances that are not components of the extracellular matrix; and (3) many members of the MMP family may not yet have been discovered. As the years have passed, all three of these predictions have come to pass. Then, in 1993, a Gordon Research Conference devoted entirely to MMPs was inaugurated in Plymouth, New Hampshire (Fig. 2.5a and b of MMPs around the dinner table). In the space of 30 years and following the identification of one enzyme with "activity" against interstitial collagen, a host of MMP family members had been identified and detailed molecular investigations into their structure and function were furiously underway. MMPs were an official topic of research (Fig. 2.4).

Almost simultaneously with these national and international meetings, and continuing into the 1990s, studies were being carried out that produced the unexpected findings about the MMP family members. As already noted, years earlier, John Jeffrey used immunoprecipitation

(a)

(b)

Figure 2.5. MMP family members seated around dinner table. "Preferred" matrix substrates are shown on dinner plates. Large hats on MMP-2 and MMP-9 correspond to fibronectin-like domains in these MMPs. MMP-7 is a truncated MMP. "R.I.P." = duplicate MMPs isolated by different laboratories. MMPs in ceiling are membrane-bound enzymes. (a) MMP family, circa 1990 (anonymous). (b) MMP family, circa 2010. (Courtesy of Dr. Gilian Murphy.)

to note distinct differences between rat and human collagenases. In 1990, we began to understand what was behind this discrepancy when rat interstitial collagenase was cloned. Sequence analysis showed only minimal identity between the rat gene and rabbit and human collagenases (MMP-1). The mystery was resolved in 1994 when the human homologue of rat interstitial collagenase was cloned.[10] Now, along with collagenase-1 (MMP-1), neutrophil collagenase (collagenase-2; MMP-8), there was yet another human collagenase: collagenase 3, or MMP-13. MMP-13 shows only partial amino-acid identity with MMP-1, indicating that it is truly a distinct enzyme.

Thus, MMP-1 appeared to be the major interstitial collagenase in rabbits and humans, while MMP-13 was the dominant interstitial collagenase in rats and mice. Concomitantly, MMP-13 expression in humans and rabbits was far less prominent. The recognition that rodents did not express *bonafide* MMP-1 has been important in studies with rats and mice where investigators tried to monitor MMP-1 expression; it was not there. Measuring MMP-13 was, and continues to be, the most appropriate target in rodents.

Nonetheless, the question remained as to whether there was a true homologue of human MMP-1 in rats and mice, and in 1994, this homologue was cloned and shown to be only a distant relative with 58% amino-acid identity. Called murine *mmp1a*, its pattern of expression was very different from *MMP-1*, occurring mostly during development. Consequently, it was essentially discarded as irrelevant to most murine/rat model systems of human disease. Recently, however, murine mmp1a has been shown to be expressed under conditions of inflammation and cancer, and thus, may be useful in some murine models of human disease.[18]

The 1990s brought yet another surprise: the identification of a membrane-bound MMP: MMP-14 or MT1-MMP.[10] Although this MMP was synthesized in the traditional latent form, the enzyme was activated intracellularly by cleavage via a prohormone convertase, furin. As a result, the enzyme was inserted into the cell membrane in activated state. Subsequently, several other membrane-bound MMPs were isolated to bring the number to six at present.

Four of these, MT1-MMP/MMP-14, MT2-MMP/MMP-15, MT3-MMP/MMP-16, and MT5-MMP/MMP-24 are anchored to the cell surface by transmembrane domains, while MT4-MMP/MMP-17 and MT6-MMP/MMP-25 are anchored to the membrane by a glycosylphosphatidylinositol (GPI) anchor, rather than by a transmembrane domain.

2.1.1 *Summary and conclusions*

To date, 28 human MMPs have been identified, with two gene duplications,[12,17,19,20] along with 23 MMP genes in mice.[20] These MMPs are expressed in nearly, if not all, types of cells, ranging from connective tissue cells (fibroblasts, endothelial cells, leukocytes, neutrophils, macrophages/monocytes, T lymphocytes, eosinophils) to most cancer cells.[10,12,20,21] Understanding the structure of cDNAs and genomic clones for the MMP family members was the first step in figuring out the mechanisms regulating MMP expression, and of critical importance, determining how this regulation contributed to normal physiology and to disease pathology.

2.2. Structure and Function of MMPs

Solving the crystal structure of MMPs was the next important step in identifying the domains of the MMP molecules that define their biochemistry and enzymatic activities.[10,12,17,20–24] The domain structures of the MMPs are shown in Fig. 2.6.[20,22] They are made up of a signal peptide, which targets these enzymes for secretion, a propeptide and a catalytic domain, both of which interact to keep the enzyme in latent form. The catalytic domain, which has about 170 amino acids, contains a Zn++ binding motif (HEXXHXXGXXH) and a conserved methionine. This domain is composed of a five-stranded beta-sheet, three alpha-helices, and bridging loops. The structural zinc ion and two to three calcium ions are required for stability and for enzymatic activity. While the overall spherical topology of MMPs is conserved among the family members, there is a groove that binds substrate. It is a narrow, well-defined pocket,

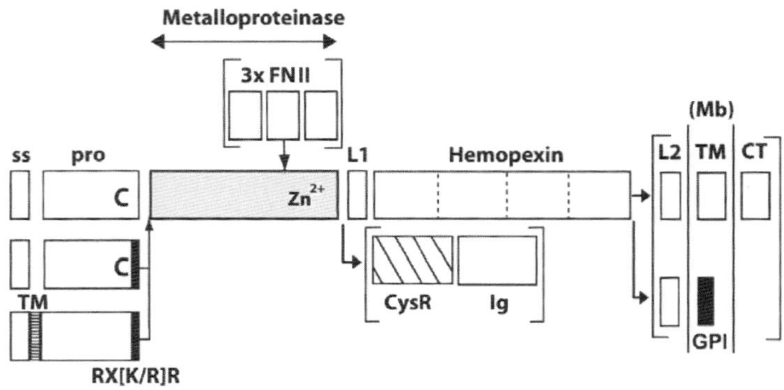

Figure 2.6. Domain structures of the MMP family. ss, signal sequence; pro, prodomain, FNII, fibronectin type II motif; L1, linker 1; L2, linker 2; Mb, plasma membrane; TM, transmembrane domain; CT, cytoplasmic tail; CysR, cysteine rich; Ig, immunoglobulin domain; GPI, glycosylphosphatidylinositol, anchor; C, cysteine. (From Murphy and Nagase, 2008.)

and variations among the amino acid residues within this pocket differentiate one MMP from another. A short linker domain joins the catalytic and hemopexin domains. The C-terminal hemopexin-like domain contains approximately 210 amino acids. It is shaped like an ellipsoidal disk with a four blade-propeller structure where each blade consists of four antiparallel beta-strands and an alpha-helix. This domain is essential for the collagenases to cleave triple helical interstitial collagens, but the catalytic domain, by itself, has proteolytic activity against other substrates.

The catalytic domains of MMP-2 and MMP-9 contain three repeats of fibronectin type II within the catalytic domain. These repeats allow the enzymes to bind to and process denatured collagen (gelatin). The hemopexin domain of MMP-2 is also needed for MT1-MMP to activate pro-MMP-2 (see the following). In contrast, the hemopexin domain is not found in MMP-7, MMP-23, and MMP-26. At least with MMP-7, this lack of hemopexin domain is associated with broad substrate specificity.

In 1990, several investigators introduced the term "cysteine switch" to describe the mechanism by which MMPs maintain enzyme latency.[10,12,20] In the proform of inactive MMPs, there is an unpaired cysteine in the highly conserved "Pro–Arg–Cys–Gly–X–Pro–Asp".[5,10] This unpaired cysteine forms a bridge with the catalytic zinc to prevent enzymatic activity. Activation occurs when this bridge is disrupted either proteolytically by a (polarized) water molecule mounting a "nucleophilic" attack or chemically by aminophenylmercuric acid (APMA).

As in other proteolytic systems, activation of MMPs can be mediated by a proteolytic cascade.[10] If it is present within the tissue microenvironment, human MMP-3 (stromelyisn-1; "activator protein") can activate human pro-collagenase. Furthermore, serine proteinases readily activate latent MMPs, illustrating cooperativity among classes of proteinases. Plasmin, which is produced by the action of a plasminogen activator on plasminogen, activates several MMPs, by cleaving the pro-domain. Other serine proteinases, such as thrombin, tryptase, and kallikrein also activate latent MMPs, thereby providing multiple pathways for eliminating latency within an *in vivo* environment. This is an important consideration when comparing *in vitro* and *in vivo* experimental systems, since the *in vivo* environment is often more complex and may have mechanisms for activating latent MMPs that are not present *in vitro*.

In 1994, another important mechanism for activating proMMP-2 mediated by MT1-MMP in a two-step process that involves TIMP-2 was described.[10,12,25] The three proteins form a complex in which TIMP-2 links the two MMPs. In this complex, MT1-MMP is inactivated due to interactions with the N-terminus of TIMP-2, while the C-terminus binds to the hemopexin domain of proMMP-2. Another MT1-MMP molecule then cleaves proMMP-2, which is partially activated. Full activation results from autocatalysis. More recently, other novel mechanisms for activating latent MMPs have been described.[25] These involve the binding of molecules such as certain glycoproteins or thiol-binding reagents to noncatalytic (exosites) within MMPs, resulting in a conformational change in the MMP molecule that may or may not result in removal of the propeptide, but that does allow enzymatic activation. Other molecular interactions that activate proMMPs include the binding of elastin and heparin, the formation of MMP-9 homo- and heterodimers through the hemopexin domain, and chondroitin sulfate proteoglycans. These interactions also mediate conformation changes, which result in intramolecular autoactivation.

2.2.1 *Summary and conclusions*

Data on the basic characteristics of MMP structure and function, mechanisms of activation of enzymatic activity, and cloning and activation have set the stage for studies on how expression of the genes encoding MMPs is regulated. Understanding the mechanisms controlling active enzyme and gene expression is essential for understanding the consequences of MMP expression in normal physiology and in diseases, as well as for discoveries of their unexpected and novel behaviors.

References

1. Ricard-Blum, S (2011). The collagen family. *Cold Spring Harbor Perspectives in Biology,* 3, a004978. http://cshperspectives.cshlp.org/content/3/1/a004978.short.
2. Di Lullo, GA, SM Sweeney, J Korkko, L Ala-Kokko and JD San Antonio (2002). Mapping the ligand-binding sites and disease-associated mutations on the most abundant protein in the human, type I collagen. *Journal of Biological Chemistry,* 277, 4223–4231.
3. Fang, M, R Jacob, O McDougal and JT Oxford (2012). Minor fibrillar collagens, variable regions alternative splicing, intrinsic disorder, and tyrosine sulfation. *Protein Cell,* 3, 419–433.
4. Manka, SW *et al.* (2012). Structural insights into triple-helical collagen cleavage by matrix metalloproteinase 1. *Proceedings of the National Academy of Sciences of the United States of America,* 109, 12461–12466.
5. Nagase, H and Y Okada (1997). Proteinases and matrix degradation. In *Textbook of Rheumatology,* WN, Kelley, S Ruddy, ED Harris and CB Sledge (eds.), 5th Ed., pp. 323–341. Philadelphia: W.B. Saunders.
6. Woessner, JF (1962). Catabolism of collagen and non-collagen protein in the rat uterus during post-partum involution. *Biochemical Journal,* 83, 304–314.

7. Gross, J and CM Lapiere (1962). Collagenolytic activity in amphibian tissues: a tissue culture assay. *Proceedings of the National Academy of Sciences of the United States of America,* 48, 1014–1022.
8. Jeffrey, JJ and J Gross (1970). Collagenase from rat uterus. Isolation and partial characterization. *Biochemistry,* 9, 268–273.
9. Nagai, Y, CM Lapiere and J Gross (1966). Tadpole collagenase. Preparation and purification. *Biochemistry,* 5, 3123–3130.
10. Brinckerhoff, CE and LM Matrisian (2002). Matrix metalloproteinases: a tail of a frog that became a prince. *Nature Reviews Molecular Cell Biology,* 3, 207–214.
11. Bauer, EA, AZ Eisen and JJ Jeffrey (1970). Immunologic relationship of a purified human skin collagenase to other human and animal collagenases. *Biochimica et Biophysica Acta,* 206(1): 152–160.
12. Murphy, G and H Nagase (2008). Progress in matrix metalloproteinase research. Molecular Aspects Medicine, 29, 290–308.
13. Liotta LA, UP Thorgeirsson, S Garbisa (1982). Role of collagenases in tumor cell invasion. *Cancer Metastasis Review,* 1(4):277–288.
14. Harper, E, K Bloch and J Gross (1971). The zymogen of tadpole collagenase. *Biochemistry,* 10, 3035–3041.
15. Nagase, H, AJ Barrett and JF Wosessner (1992). Nomenclature and glossary of the matrix metalloproteinases. *Matrix Supplement,* 1, 421–424.
16. Birkedal-Hansen, H (1988). From tadpole collagenase to a family of matrix metalloproteinases. *Journal of Oral Pathology,* 17, 445–451.
17. Iyer, RP, NL Patterson, GB Fields and ML Lindsey (2012). The history of matrix metalloproteinases: milestones, myths, and misperceptions. *American Journal of Physiology Heart Circulatory Physiological,* 303, H919–H930.
18. Foley, CJ and A Kuliopulos (2014). Mouse matrix metalloprotease-1a (Mmp1a) gives new insight into MMP function. *Journal of Cellular Physiology,* 229, 1875–1880. doi:10.1002/jcp.24650.
19. Puente, XS., LM Sánchez, CM Overall and C López-Otín (2003). Human and mouse proteases: a comparative genomic approach. *Nature Reviews Genetics,* 4, 544–558.
20. Fanjul-Fernández, M, AR Folgueras, S Cabrera and C López-Otín (2010). Matrix metalloproteinases: evolution, gene regulation and functional analysis in mouse models. *Biochimica Biophysica Acta,* 1803, 3–19.
21. Rivera, S, M Khrestchatisky, L Kaczmarek, GA Rosenberg and DM Jaworski (2010). Metzincin proteases and their inhibitors, foes or

friends in nervous system physiology? *Journal of Neuroscience,* 30, 15337–15357.

22. Nagase, H and JF Woessner (1999). Matrix metalloproteinases. *Journal of Biological Chemistry,* 274, 21491–21494.
23. Ra, HJ and WC Parks (2007). Control of matrix metalloproteinase catalytic activity. *Matrix Biology,* 26, 587–596.
24. Gomis-Rüth, FX (2009). Catalytic domain architecture of metzincin metalloproteases. *Journal of Biological Chemistry,* 284, 15353–15357.
25. Hadler-Olsen, E, B Fadnes, I Sylte, L Uhlin-Hansen and JO Winberg (2011). Regulation of matrix metalloproteinase activity in health and disease. *FEBS Journal,* 278, 28–45.

3. Regulation of Matrix Metalloproteinase (MMP) Gene Expression

Constance E. Brinckerhoff

3.1 Introduction

The previous chapters have discussed mechanisms by which the biochemical activity of matrix metalloproteinases (MMPs) is regulated, either by proteolytic activation of latent forms and/or by blocking enzymatic activity with endogenous inhibitors. Although both are important in controlling matrix remodeling, the mechanisms controlling expression of MMP genes are dominant in determining the levels of MMPs found in normal physiology and in disease pathology. Three mechanisms regulate the expression of genes.[1] The first is *transcription*, that is, the process by which a gene (double-stranded DNA that encodes a functional protein) is transcribed into a single strand of mRNA by a complex assortment of proteins (transcription factors) that bind to specific sequences in the control region ("promoter") of a gene. Transcription first produces heterogeneous nuclear ribonucleic acid (hnRNA), a virtual copy of the genomic DNA, which contains introns and exons. The introns are cut/spliced out, leaving "mature mRNA", comprised of (1) a 5′-untranslated region (5′-UTR), (2) the coding region that will be translated into protein, and (3) a 3′-untranslated region (3′-UTR) (Fig. 3.1). The second mechanism is *posttranscriptional modification* and represents the processes by which levels of mRNA are

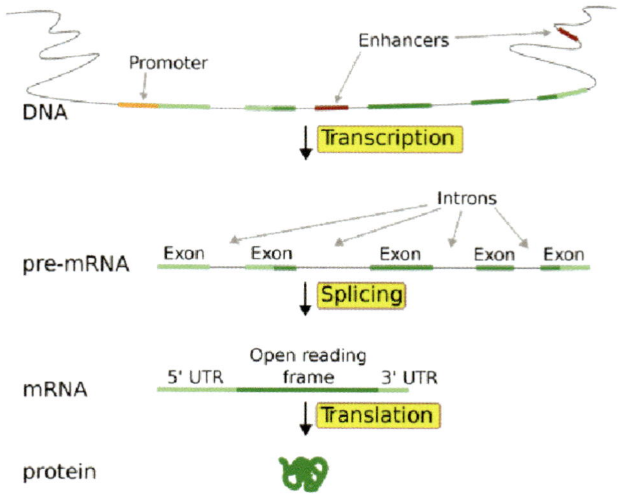

Figure 3.1. The basics of gene expression. A gene is composed of DNA, which is transcribed into mRNA and then translated into protein. The protein is eventually expressed by the cell where it carries out specific function(s). (From Wikipedia.)

modulated by proteins that bind to the 3′-UTR and prevent or facilitate its degradation. It is, therefore, important to emphasize that amount of a specific mRNA that encodes a particular gene is the net sum of transcription and posttranscriptional modification; mRNA levels, *per se*, are not an accurate reflection of transcription, a point that is often overlooked. The half-life of mRNAs can be stabilized by proteins that bind to the 3′-UTR and retard mRNA degradation. Alternatively, micro-RNAs (miRNAs), small double-stranded RNAs of about 20 nucleotides, bind to the 3′-UTR of specific mRNAs and facilitate their degradation. Lastly, *epigenetic mechanisms* influence the levels of expression of a gene by chromatin remodeling through acetylation and methylation of histones. These epigenetic modifications of genes can either increase or decrease access of transcription factors to DNA sequences of particular genes, thereby resulting in increased or decreased transcription.

In normal physiology, expression of MMPs is generally low as these enzymes maintain homeostasis of connective tissues (Table 3.1).[2] However, under certain normal but temporary physiological condi-

Table 3.1. Expression of MMPs.

Physiology	Pathology
• Morphogenesis	• Rheumatoid/Osteo-arthritis
• Reproduction	• Tumor invasion/metastasis
• Wound healing	• Cardiovascular disease
	• Chronic wounds
• Low MMP levels	• High MMP levels

tions, such as acute wound healing, stages of embryonic development, angiogenesis, and uterine resorption following pregnancy, there are localized spatial and temporal spikes in the expression of selected MMPs. In contrast, in many diseases, such as rheumatoid and osteoarthritis, cancer, cardiovascular disease, and chronic wounds, there is aberrantly high MMP expression, which is only minimally or poorly controlled (Table 3.1). This dysregulated MMP expression contributes substantially to disease pathology. Therefore, understanding the basic molecular mechanisms that control expression of each MMP in health and disease, and in a tissue-specific manner, has enormous implications for understanding the molecular pathogenesis of disease and for devising therapeutic strategies to subvert this overexpression.

3.2 Transcriptional Regulation

Following almost immediately on the heels of the era of cloning and sequencing of genes, molecular biology began to focus on the promoter region, that is, the DNA 5′ (upstream) of the coding sequences (Fig. 3.2).[3–9] Curiously, the promoter of the human collagenase gene was an initial object of intense investigation, not because of any inherent interest in understanding how expression of the collagenase gene was controlled, *per se*, but rather because the promoter contained a sequence of DNA, which came to be known as the activator protein-1 (AP-1) site. Again using the primitive technology that was available at the time, genomic clones for human collagenase were isolated and the adjacent 5′ promoter region was cloned.[10] Also at

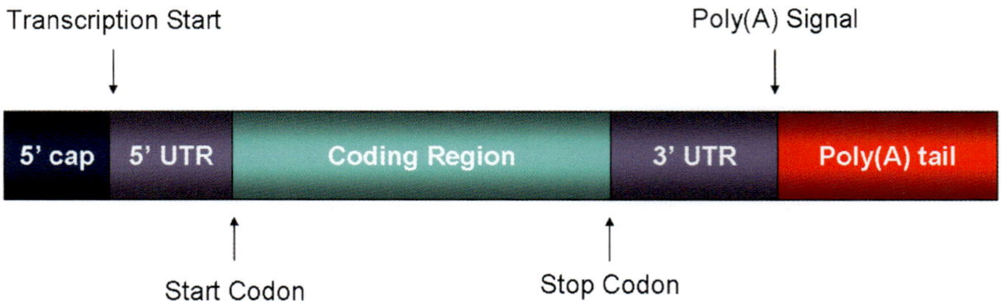

Figure 3.2. Structure of mRNA. (From Wikipedia.)

that time, there was considerable interest in the plant-derived compound, phorbol ester, 12-O-tetradecanoylphorbol-13-acetate (TPA) (also known as phorbol myristate acetate [PMA]), a potent transcriptional activator, with both inflammatory and tumor promoting activities. Because of its structural similarity to diacylglycerol and its ability to activate protein kinase C (PKC),[4] it was a popular experimental tool, but has now been largely abandoned because it is not physiologically relevant. Perhaps more importantly, naturally occurring physiologically active growth factors and inflammatory cytokines were found to be potent transcriptional activators.

However, by using phorbol esters as an experimental tool and by constructing mutations in the 5′ control (promoter) region of the collagenase gene and transfecting the mutant constructs into human or murine cells, a region upstream of the transcription start site was termed as the "phorbol ester response element".[10] This "*cis*-acting sequence"[1] is located in a region of DNA at approximately –73 to –42 bp, and contains the consensus sequence, 5′-ATGAGTCAT-3′.[11] This element was absolutely required for the induction of transcription by TPA. It behaved as a classical enhancer: increasing transcription of a heterologous herpes simplex virus thymidine kinase promoter and acting in a position- and orientation-independent manner.[10–12] In addition, the identity of the proteins ("trans-acting factors") that bound to the AP-1 site was identified as members of the Fos and Jun families of transcription factors.[2,3] This work served as an excellent cornerstone for subsequent studies that identified AP-1 sites in other MMPs (and other genes as well).[3,11]

Indeed, AP-1 sites were present in many other genes, and many of us in different labs were working simultaneously to identify important regulatory sequences, including AP-1, in several other MMP genes, for example, MMP-3 and MMP-9.[4–7] It also became clear that these genes were often coordinately regulated with MMP-1 in response to stimuli. Before long, however, another *cis*-acting sequence in the MMP-1 promoter was found to be important in gene expression: the erythroblastosis twenty six site ("ETS"; also known as polyomavirus enhancer activator 3 [PEA3]), which lies in close proximity to the AP-1 enhancer. A classic study with careful point mutations of single nucleotides demonstrated that an ETS site, located at approximately −82 to −89 bp, with a core sequence of 5′-GGAA-3′ cooperated with AP-1 to regulate transcription.[13] Quickly, the story became more complicated: the proximal AP-1 site, even in cooperation with ETS, although necessary, was not sufficient to regulate MMP gene expression. Subsequent studies showed the importance of cooperation with several upstream elements in regulating gene expression of various MMP promoters.[4–7] Eventually, numerous DNA sequence motifs were identified in the promoters of many genes, including MMPs, which bound a myriad of transcription factors to modulate gene expression, resulting in both transcriptional activation and repression (Fig. 3.3). Multiple sites for the same element may be found within a given promoter, and their identification along with the transcription factors that bound to them, solidified the concept of cooperativity among *cis*-acting promoter sequences. It is now recognized and accepted that no single sequence could single-handedly maximally drive (or silence) transcription.

Furthermore, sequences regulating MMP gene expression may be found quite distal to the proximal promoter, perhaps brought into close contact with the transcriptional machinery because of the configuration of native chromatin. Studies with the MMP-13 gene have identified an interleukin-1β (IL-1β) response element in the MMP-13 promoter 20 kb upstream of the transcription start site.[14] In other studies, a new role for the retinoblastoma (Rb)-E2F transcriptional regulatory pathway, usually associated with cell-cycle regulation, has been reported.[15] Although E2F binding sites were

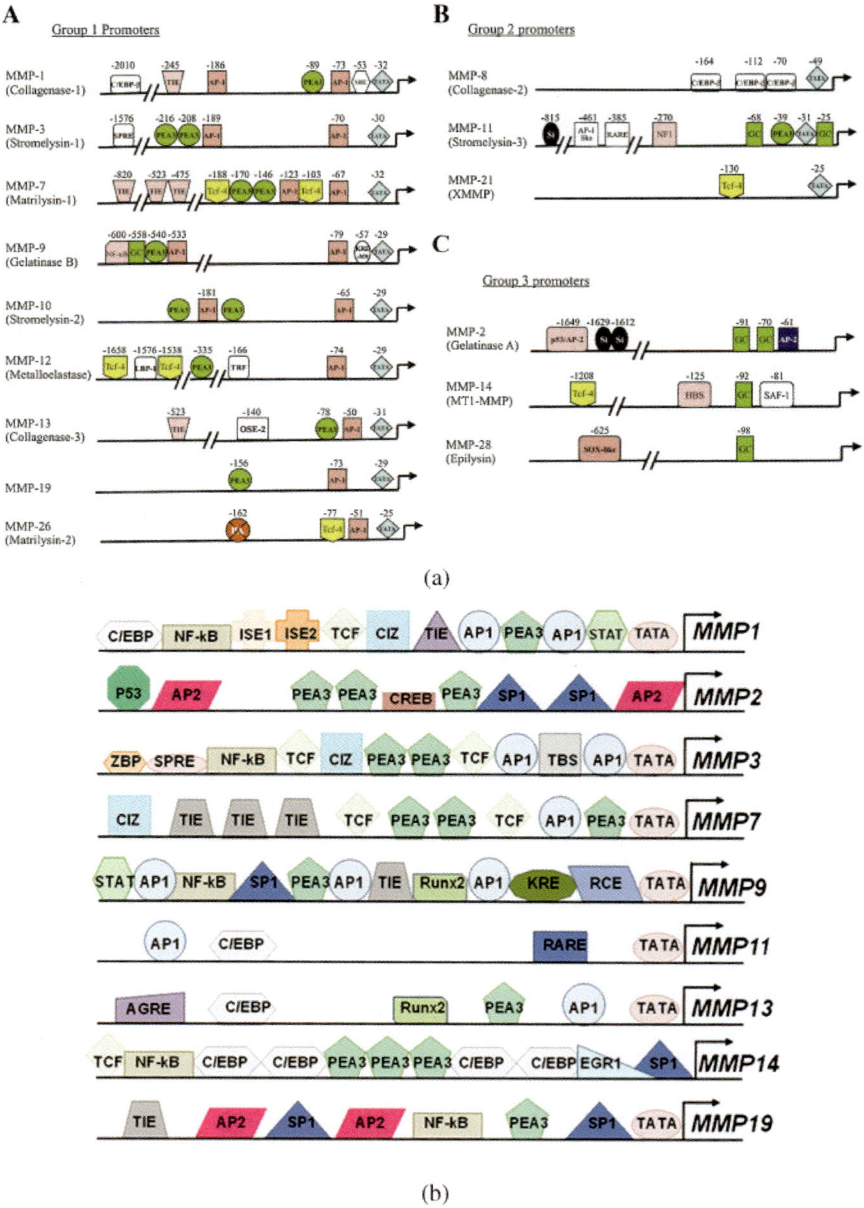

Figure 3.3. *Cis*-elements in human MMP promoters. AP-1, activator protein-1; AP-2, activator protein-2; C/EBP-b, CCAAT/enhancer binding protein-b; GC, Sp-1-binding site; HBS, HIF-binding site; KRE-M9, keratinocyte differentiation factor-1

predicted on all 23 human MMP gene promoters, three MMPs (MMP-9, MMP-14, and MMP-15) are overexpressed in non-small-cell lung cancer and are regulated by the RB-E2F pathway. These data suggest that E2F transcription factors may contribute to metastasis by increasing MMP expression, thereby enhancing the invasive behavior of tumors along with increasing cell proliferation. Finally, once thought to be expressed only during embryogenesis, the murine homologue of human MMP-1, mmp1a, has now been demonstrated to be expressed in response to inflammatory stimuli.[16] Induction of septic shock by administration of lipopolysaccharide (LPS) results in MMP-1a release from murine endothelial cells. Mmp1a has also been elevated in several murine tumor models of cancer, suggesting that mouse models may be used to understand MMP-1 expression within the context of an animal model.

Based on the organization of their *cis*-elements, the MMP promoters have been grouped into three categories (Fig. 3.3).[7] In the first group (A), which includes most MMP promoters, the organization is remarkably similar. They share many *cis*-acting sequences, which explains how they may be coordinately regulated. The promoters in this group contain a TATA box at about –30 bp (relative to the transcription start site), which recruits a core of basic transcription factors and assures stable transcriptional initiation from a particular site. Importantly, there is also the proximal AP-1 binding site at about –70 bp, as well as an adjacent upstream PEA3-binding site, which cooperates with the proximal AP-1 site.[5]

The second group (B) of MMP promoters contains MMP-8, -11, and -21. They also harbor a TATA box, but lack a proximal AP-1

Figure 3.3. (*Continued*) responsive element-4; LBP-1, leader-binding protein; NF1, nuclear factor-1; NF-κB, nuclear factor-κB; OSE-2, osteoblast-specific element-2; p53/AP-2, p53/AP-2 composite binding site; PA, polyadenylation signal; PEA3, polyoma enhancer A binding protein-3; RARE, retinoic acid (RA) responsive element; SBE, STAT-binding element; Si, silencer sequence; SPRE, stromelysin-1 PDGF-responsive element; TATA, TATA box; Tcf-4, T-cell factor-4/b-catenin-binding site; TIE, TGF-β inhibitory element; TRF, octamer-binding protein; SAF-1, serum amyloid A activating factor-1. (From Yan and Boyd, 2006.)

site. Without this site, their regulation is distinct from the first group and depends on upstream sequences for the control of gene expression. The third group (C) of promoters includes MMP-2, -14, and -12. There is no TATA box and transcription from these promoters can be initiated at multiple sites. Expression of these MMPs is often "constitutive", occurring with only minimal regulation by growth factors or cytokines. However, the Sp-1 family of transcription factors, which is commonly found in many cells, binds to a proximal GC box and may influence expression of these MMPs.

To summarize, transcription of genes belonging to the MMP family is regulated by several distinct promoter elements and by the proteins that bind to them. Interestingly, and of some importance, the regulation of several MMPs that are functionally similar (e.g., the collagenases MMP-1 and MMP-8, and the gelatinases MMP-2 and MMP-9) is quite different. This suggests that it may be biologically important to have differential mechanisms regulating the expression of genes with the same functions.

3.3 Signal Transduction and the Induction of MMP Gene Expression

No discussion on transcriptional regulation can occur without considering "signal transduction", the process by which an extracellular molecule, such as a growth factor, cytokine, or hormone, binds to a specific receptor on the cell surface and triggers a biochemical cascade of events inside the cell. This cascade eventually ends in the nucleus, where it creates a response by affecting expression of target genes.[7,17–19] Depending on the cell and the signal, these responses result in changes in cell metabolism, shape, gene expression, or proliferation.

The field of signal transduction was "born" at about the same time promoter analysis began, as cell membrane receptors for hormone and neurotransmitters were found to form complexes with guanosine triphosphate (GTP)-binding proteins.[18,20] The work eventually led to Nobel Prizes for Martin Rodbell and Alfred Gilman. However, the focus was at the cell membrane, while concomitantly, promoter analysis was confined to the nucleus.[3,5,7,11,13] As knowledge in each field progressed, eventually the two areas of study coalesced, and as Dr. Rodbell stated in his Nobel address: "Those of us attempting to view the living process at the larger cellular level will merge with our assemblages of ideas and experiences [to find] a bright new era in scientific discovery".[20]

Signal transduction has become a very complicated area of investigation, and numerous signaling pathways are present within most cells (Figs. 3.4 and 3.5).[19] The signal is transmitted by a particular pathway and is perpetuated through the cell by a cascade of phosphorylation events, which culminate in the activation of transcription factors in the nucleus, where they bind to promoter elements to increase gene expression. There is often considerable "cross talk" as well as some functional redundancy since the different pathways target the same promoter elements in several genes. Since many ligands can activate the same intracellular signaling

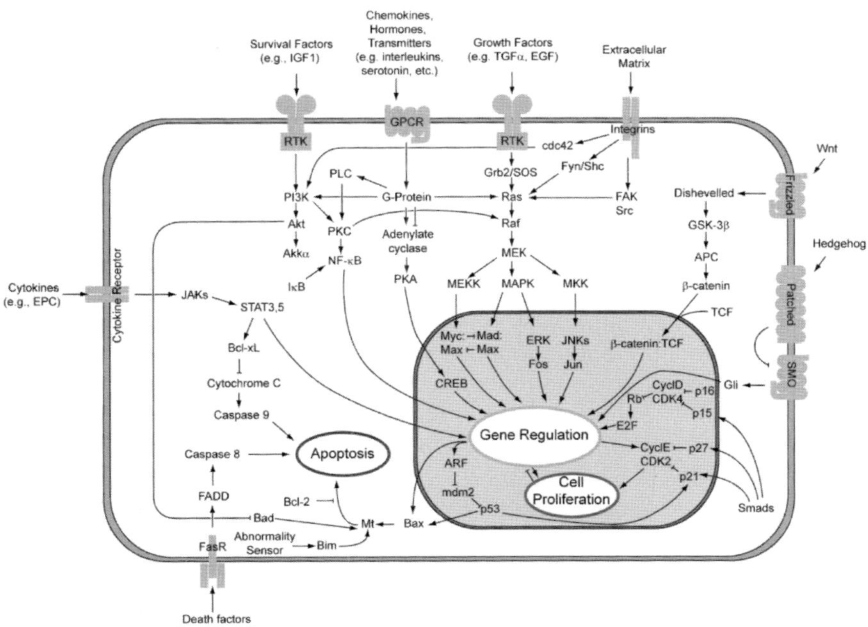

Figure 3.4. Overview of signal transduction pathways. (From Wikipedia.)

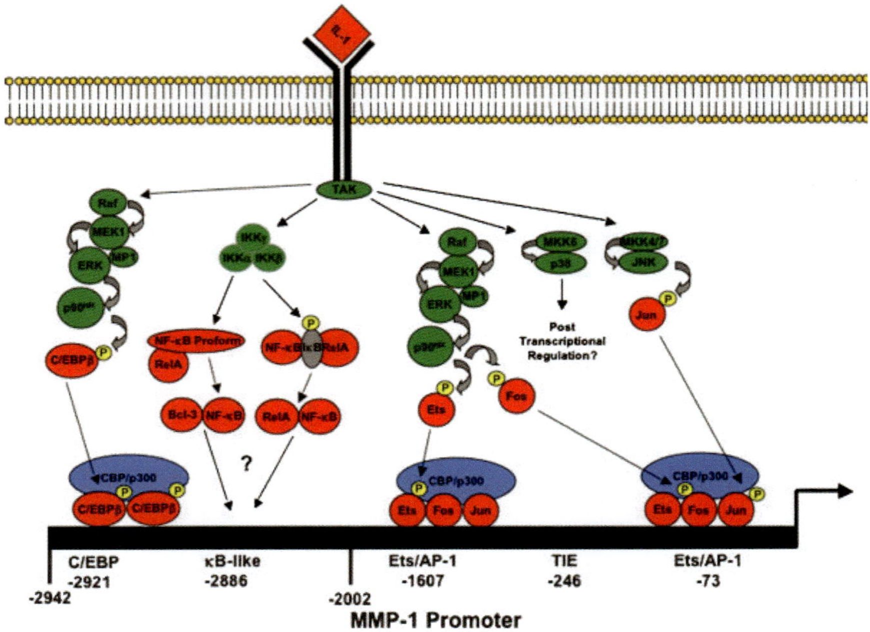

Figure 3.5. Example of a signaling pathway that regulates the MMP-1 promoter. The cytokine, IL-1, binding stimulates the TGF-activated kinase (TAK), which activates the MAPK (ERK, p38, and JNK) and NF-κB pathways. The ERK pathway activates the kinase $p90^{rsk}$, which phosphorylates and activates essential transcription factors, including C/EBPb, Ets, and Fos. Activation of the JNK pathway results in phosphorylation and activation of Jun, which promotes transcription from AP-1 elements. Phosphorylation of these transcription factors promotes recruitment of transcriptional coactivator proteins such as CBP/p300. The p38 targets in the MMP-1 promoter are not known, but this pathway may also contribute to mRNA stabilization. IL-1 activation of the NF-κB pathways results in IκB degradation and nuclear localization of RelA-containing dimers, as well as processing of latent NF-κB. Latent NF-κB can act as IκB proteins and sequester RelA in the cytoplasm. Processed NF-κB proteins lack transactivation domains but transcription can be promoted through an association with the IκB family member protein Bcl-3. The composition of NF-κB proteins that bind to the MMP-1 promoter is not completely defined. (From Vincenti and Brinckerhoff, 2007.)

pathways, how do signal- and cell-specific pathways activate MMP transcription? What are these signal transduction pathways and how do they "transmit" their signal? While this is a complex issue, it is apparent that the integration of multiple pathways and the recruitment of multiple transcription factors are somehow integrated to result in this specificity. It is important to emphasize the complexity of these pathways, which often recruit cofactors/coactivators to assist in accurately perpetuating the signal from the cell membrane to the nucleus.

1. The *mitogen activated protein kinases (MAPKs)* are major mediators of signal transduction for many genes, including MMPs. They belong to the family of enzymes called kinases that phosphorylate proteins on serine and threonine residues. Once phosphorylated, these kinases activate transcription factors and RNA-modifying proteins in response to inflammatory stimuli, growth factors, and cell–matrix interactions. The MAPKs can be classified into three groups: (1) the extracellular-regulated kinases (ERKs); (2) the stress-activated or p38 protein kinases; and (3) the c-Jun N-terminal kinases (JNKs). A host of cytokines, including IL-1β, tumor necrosis factor-α (TNFα), Oncostatin M, and receptor activator of nuclear factor kappa-B ligand (RANKL), microbial products such as LPS, and a wide variety of growth factors, such as epidermal growth factor (EGF), nerve growth factor (NGF), basic fibroblast growth factor (bFGF), vascular endothelial growth factor (VEGF), platelet-derived growth factor (PDGF), and transforming growth factor-beta (TGF-β), activate MAPK pathways and increase transcription by targeting *cis*-acting sequences within the MMP promoters.[7,19] Since many extracellular signals (i.e., growth factors and cytokines) can activate one or more MAPK pathways, the ability of a MAPK pathway to activate a particular gene is linked to its ability to target specific transcription factors that, in turn, activate a particular MMP.[7,19]

For example, inflammatory cytokines and growth factors are potent activators of the ERK pathway in both tumor cells and in stromal cells (Figs. 3.4 and 3.5).[19] When these ligands bind to their receptors, which are transmembrane tyrosine kinases, the kinases phosphorylate and recruit "adaptor proteins" (e.g., Shc, Grb2, and

Sos). These adaptors then recruit the G protein, Ras, to the cell surface, followed by the sequential activation of Raf, MEK1, and ERK1/2. Since activation of the ERK pathway can activate several MMPs at the same time, including MMP-1, MMP-3, MMP-7, MMP-9, and MMP-14, along with many other genes, ERK signaling has a major role in signal transduction.

In some cancers, mutations result in the constitutive activation of the ERK pathway, bypassing the need for ligand stimulation. The EGF receptor (EGFR) is mutated in several cancers, including breast and lung, and as a result, it signals constitutively even in the absence of its ligand, EGF.[7,19] Similarly, mutations in melanoma, particularly the $BRAF^{V600E}$ mutation, constitutively activate ERK and MMP-1 transcription.[21] The constitutive phosphorylation of ERK transactivates the MMP-1 promoter through AP-1 and Ets proteins, which bind to several sites located in the MMP-1 promoter, resulting in high levels of MMP-1 expression and a highly invasive phenotype in these cancer cells.[22,23]

The p38 protein kinases are activated primarily in response to inflammation and innate immunity, and they, too, regulate numerous MMP genes: MMP-1, MMP-3, and MMP-13. However, despite the potential to induce several MMPs, p38 signaling provides an excellent example of tissue-specific targets. The MMP-13 promoter is activated by p38 in chondrocytes through recruitment of AP-1 proteins to the nearly ubiquitous AP-1 site in the proximal promoter, but only in cooperation with transcription factor, RUNX-2. Since RUNX-2 is a chondrocyte/osteoblast-specific transcription factor, its cooperativity with the more promiscuous AP-1 nicely demonstrates a mechanism for the tissue-specific regulation of MMP transcription.

Lastly, the Jun amino-terminal kinases (JNK) family is comprised of three genes (isoforms JNK 1,2,3), which are also activated in response to inflammatory stimuli. Once JNK is phosphorylated, it activates the transcription factors ATF2, Elk-1, and c-Jun. The JNK pathway is required for cytokine induction of both MMP-1 and MMP-13 in synovial fibroblasts and for MMP-13 expression in chondrocytes. However, induction of MMP-1 by IL-1β in chondrocytes does not require JNK pathway, again demonstrating the cell-type specificity of signal transduction pathways and gene expression. The differential activation of two human interstitial collagenases, MMP-1 and MMP-13, is striking and provides a way to achieve tissue specificity and to avoid unnecessary redundancy of gene expression.

2. TGF-β belongs to a ubiquitous superfamily of cytokines with diverse effects on many genes in many different cells and tissues.[24] Because the effects of the TGF-β family of proteins are contextual and can be so varied, it is almost impossible to assign definitive roles to these proteins. This superfamily includes TGF-β 1, 2, and 3 and the bone morphogenetic proteins (BMPs). All are critical modulators of connective tissue remodeling. TGF-β is a major inducer of collagen synthesis, thereby immediately giving this cytokine a prominent role in ECM metabolism, with varied effects in health and disease.[19,24,25]

The cellular receptors for TGF-β are serine/threonine kinases. These kinases must interact with a small family of intracellular coregulatory proteins called Smads, which can enhance or inhibit TGF-β-mediated gene expression. There are three subgroups of Smads, based on their different roles in TGF-β family signal transduction: R-Smads (receptor-regulated), Co-Smads (common partner), and I-Smads (inhibitory). In general, activation of TGF-β receptors leads to the phosphorylation of R-Smads (Smads1, 2, 3, 5, and 8). This is followed by recruitment of the Co-Smad (Smad-4) into an R-Smad/Co-Smad complex that translocates to the nucleus to regulate gene expression. I-Smads (Smad6 and Smad7) antagonize this signaling pathway by preventing R-Smads from interacting with the receptor. The combination of TGF-β family members, their receptors, and the assortment of coregulatory Smads results in intricate networks of regulation for many genes including MMPs. Not surprisingly, then, TGF-β and their Smads can have both positive and negative effects on MMP gene transcription.[19]

3.4 Early Response Genes

Many MMP family members have the same *cis*-acting sequences in their promoters, and consequently, they are often coinduced in response to inflammatory cytokines and growth factors, or corepressed in response to glucocorticoids or retinoids.[4,7–9,19,26] Responses to these inducing and repressing agents occur early, usually within hours, and before there is an increase in hnRNA as a result of increased transcription and this suggests that MMP promoters are downstream targets of "early response genes". These are genes that function as signaling intermediates, which phosphorylate different transcription factors, when then bind to sequences within MMPs promoters. These signaling intermediates include signal transduction kinases (i.e., the MAPKs), the nuclear factor-kappa B (NF-κB), signal transducer and activator of transcription (STAT), and the Smad family of proteins. Blocking the signaling pathways with antibodies or small molecule chemical inhibitors prevents the synthesis of some downstream mediators. As a result, transcription factors can be sequestered and/or their binding to DNA prevented. Consequently, MMP gene expression is suppressed.[4,8,9,19] Thus, along with mediators of signal/transduction pathways, transcription factors can also be early response genes, since their expression is often increased when their promoters are targets of these signaling pathways. These transcription factors must increase before transcription of downstream targets, such as MMPs, can be increased (or decreased). Some of the most common pathways and the transcription factors that increase MMP expression are listed here.

1. *AP-1 proteins*. Most MMP promoters contain an AP-1 site in the proximal promoter (~–70 bp), located close to a typical TATA box, and the transcription factors that bind to this site are often referred to as "AP-1 proteins". Members of the Fos and Jun families of transcription factors are AP-1 proteins and include proteins such as c-jun, jun-B, jun-D, c-fos, Fos-B, Fra-1, and Fra-2, all of which dimerize in various combinations. Variations in binding partners can influence the potency of gene expression.

Jun is always an obligate member of these dimers, and although Jun–Jun homodimers can modestly activate MMP transcription, transactivation is more efficiently accomplished by heterodimers formed from among the Jun/Fos family of proteins.

2. *PEA3 also known as the ETS family of proteins.* PEA3 sites bind members of the Ets family of oncoproteins, for example, Ets1, Ets2, Elf, Erg, Elk. Although PEA3/ETS sites may be found throughout MMP promoters, the principal site is located adjacent to the proximal AP-1 site, and the two sites cooperate to promote MMP production in stromal cells and in cancer cells. Transcription factors binding to PEA3/ETS sites form complexes with other transcription factors, such as AP-1 proteins. Furthermore, since PEA3 sites bind a wide variety of Ets protein family members, these proteins provide specificity of gene expression. Thus, these functional interactions between Ets and AP-1 factors demonstrate that MMP gene expression may be specifically regulated under certain conditions, either in normal physiology and/or in pathological situations, such as inflammatory conditions and tumor cell growth and invasion.
3. *NF-κB proteins.* Several growth factors and cytokines activate the NF-κB signaling pathway, leading to an increase in MMP gene expression, often in response to inflammatory conditions, such as arthritis, and in numerous cancers. When it is inactive, NF-κB is in the cytosol, complexed with the inhibitory protein, IκBα. When inflammatory cytokines engage their membrane receptors, a signaling cascade is initiated, which first involves the activation of the enzyme IκB kinase (IKK). IKK then phosphorylates the IκBα protein, which is ubiquitinated and then dissociates from NF-κB. IκBα is degraded by proteasomes. Activated NF-κB then translocates to the nucleus where it binds to specific NF-κB response elements. The NF-κB family of transcription factors includes several proteins: NF-κB1 and 2, RelA, c-Rel, and Rel-B, all of which can interact with other proteins to increase MMP expression. Consequently, NF-κB is a common mechanism used to activate MMP gene expression.

4. *Signal transducers and activators of transcription* (STAT). The family of STAT proteins resides in the cytoplasm and translocate to the nucleus after they are phosphorylated on tyrosine and, after they dimerize, they bind to STAT response elements in the promoter. They often interact with several different transcription factors to mediate gene-specific expression. For example, EGF increases MMP-1 transcription through STAT-3 signaling and by stimulating the binding of c-Jun to the AP-1 site, which is located near the STAT binding site. However, activating STAT proteins does not always increase MMP gene expression; in response to certain stimuli, STAT proteins can sequester transcriptional coactivators and prevent their ability to binding to *cis*-sequences on the promoters.

In summary, these are only a few of the transcription factors that regulate MMP gene expression. While these represent some of the more prevalent mechanisms, there are many others and all together, they contribute to the complex and subtle mechanisms that control MMP transcription in a tissue-specific manner under various physiological and pathological conditions.

3.5 Novel Mechanisms Regulating MMP Transcription (Polymorphisms)

Single-nucleotide polymorphisms (SNPs) are not mutations. Rather, they are commonly found variations in the DNA sequence of genes. While these variations can be located anywhere in a gene, when they are in the promoter region, they may affect gene regulation by changing interactions between transcription factors and sequences in the DNA, resulting in either increased or decreased transcription. Importantly, they may be associated with increased susceptibility to different diseases and with their prognosis.[9] These polymorphisms arise from nucleoside insertions, substitutions, or microsatellites, and a number of SNPs have been described in the promoters of MMPs.[6–9,19]

In the MMP-1 promoter, a polymorphism resulting from an insertion of a guanosine at position –1607 bp creates the sequence, 5′-GGAA-3′ (vs. 5′-GAA-3′), which is a consensus binding site for the Ets family of transcription factors. This "2G allele" is adjacent to an AP-1 site at –1603 bp and the two sites cooperate to enhance MMP-1 transcription (Fig. 3.6a and b).[6,27,28] The 2G allele is associated with increased promoter activity, and with a poorer prognosis in several cancers,[7–9,19] loosening of total hip replacements,[8] fibrotic disorders and cirrhosis.[9]

An insertion in the promoter of MMP-3 generates the "6A allele". Located at –1171 bp, this allele enhances the affinity of DNA for the transcription factor, ZBP-89, which is a transcriptional repressor, and thus, compared to the 5A allele, this allele decreases MMP-3 expression. The 6A allele may have protective roles in lung, oral, and breast cancer, whereas the 5A MMP3 allele is linked to increased risk for these cancers.[7–9] However, other studies on the 6A/6A genotype have been associated with poorer outcomes in rheumatoid arthritis and atherosclerotic disease.[9]

Substitutions with three nucleosides generate polymorphisms in the MMP-2 promoter. In two of them, C to T transitions are located at –735 and –1306 bp, respectively. The T allele abolishes an Sp1 binding site, thereby reducing transcription.[7–9] The third SNP, at –1576 bp, is a G to A transition, which prevents the binding of estrogen receptor to the promoter, is associated with reduced transcription in response to estrogen in some cells.

In the MMP-7 promoter, two SNPs have been found: an A to G substitution at –181 bp and a C to T substitution at –153 bp, both of which increase transcription.[9] The –181 G allele may be associated with increased susceptibility to several cancers.[9] Three substitutions have been described for MMP-8: –799C/T, –381A/G, and +17C/G. The G allele at +17 C/G SNP is associated with decreased risk for lung cancer, and the –799 T lowers susceptibility to metastasis and is linked to better survival in breast cancer patients.[9] These findings agree with the intriguing concept MMP-8 functions as a tumor suppressor[9] (see Chap. 4). Finally, the A to G substitution in the MMP-12 promoter reduces gene expression because it is directly

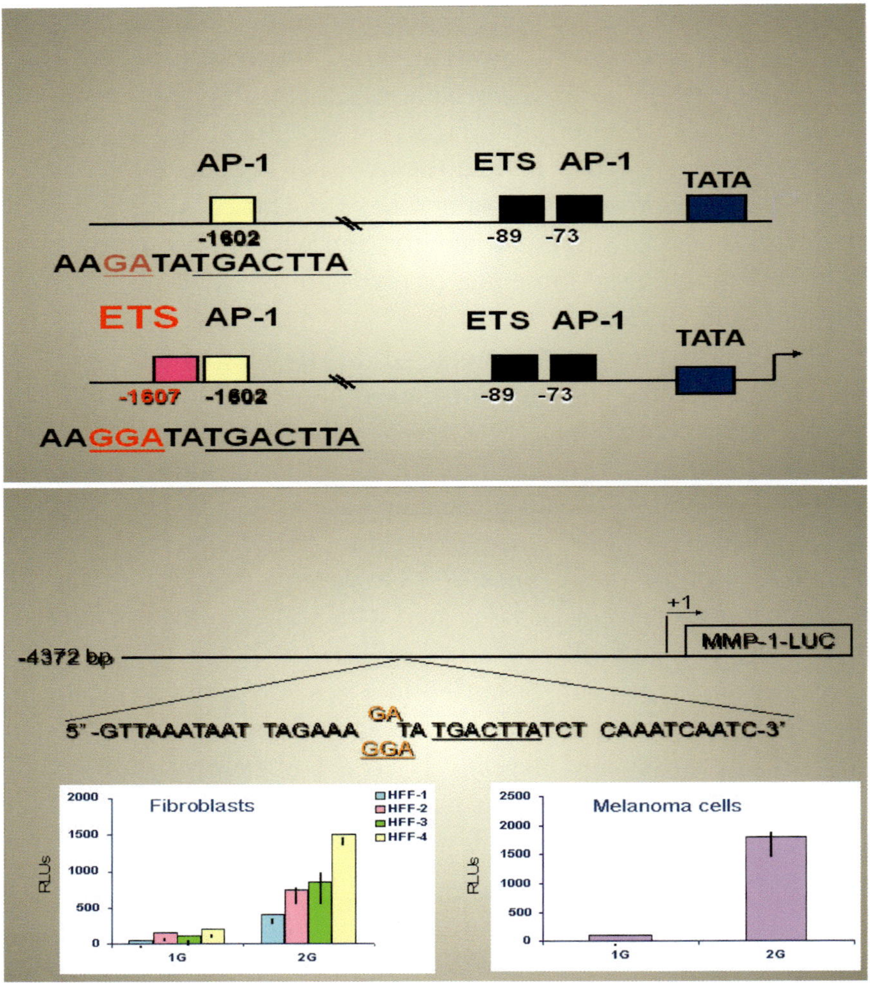

Figure 3.6. Transcriptional effects of the MMP-1 SNP at –1607 bp in human foreskin fibroblasts (HFFs) or A2058 human melanoma cells. (a) A segment of the MMP-1 promoter containing the SNP is shown. The ETS-SNP and the AP-1 sites are underlined. (b) Basal transcription of the human MMP-1 promoter with either 1 or 2Gs located at position –1607 bp in HFF cells and A2058 melanoma cells. The nucleotide sequence for the MMP-1 promoter clone has been deposited in the GenBank database under GenBank Accession Number AF023338. (Adapted from Rutter *et al.*, 1998.)

adjacent to the AP-1-binding site and reduces the binding affinity for AP-1 proteins.[7]

The MMP-9 promoter has two types of polymorphisms.[7–9] First, there is a C to T substitution at –1562 bp, which results in higher levels of gene expression. Although the identity of the transcription factor(s) responsible for the increase is not known, the T allele has been linked to increased susceptibility for cardiovascular disease.[9] However, the most noteworthy polymorphism in the MMP-9 promoter is the variable number (from 14 to 24) of microsatellite cytosine/adenine (CA) dinucleotide repeats. The repeats are located immediately adjacent to the proximal AP-1-binding site. Therefore, it has been suggested that the varying numbers of CA repeats augments the transcriptional activity of this AP-1 site,[7–9] and that the increase in MMP-9 expression is proportionally linked to the number of CA repeats. Further, these CA microsatellites have been associated with a higher risk of invasive bladder cancer and with increased susceptibility to atherosclerosis, multiple sclerosis, aneurysms, and age-related macular degeneration,[8,9] all of which suggest important functional consequences of this polymorphism.

3.6 "Targeted" Repression of MMP Transcription

Both vitamin A analogues (all-trans-RA and synthetic retinoids) and glucocorticoid hormones repress MMP mRNA and protein.[4,8,26,29] Retinoic acid receptors (RARs α, β, and γ) bind the naturally occurring isomers all-trans and 13-*cis* RA, while retinoid X receptors (RXRs α, β, and γ) bind only 9-*cis* RA. Synthetic retinoids can be designed to bind to certain RARs or RXRs, thereby hoping to achieve more specificity with target genes. RARs and RXRs heterodimerize and modulate gene expression by binding to the half-site consensus sequence, 5′-AGGTCA-3′, which is usually found in direct repeats (DRs), spaced from 1 to 5 nucleotides apart (Fig. 3.7).[4]

In contrast, glucocorticoids receptors bind to specific steroid response elements with a palindromic sequence 5′-AGAACAnnn TGTTCT-3′. However, steroid response elements are not usually

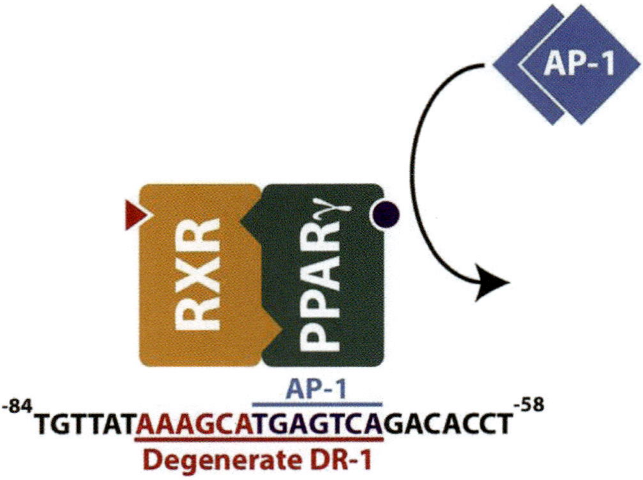

Figure 3.7. Proposed steric hindrance model for inhibiting MMP-1. Overlapping degenerate (Direct Repeat) DR-1 and AP-1 binding sites in promoter DNA. Numbers show distance from transcriptional start site in the rabbit MMP-1 gene. In the model, IL-1b induces AP-1 proteins, which bind to their cognate element and increase transcription. However, if the peroxisome proliferator activated receptor (PPARγ) ligand, rosiglitazone, or the RXR ligand, LG268, is present, the liganded RXR:PPARγ dimer binds to the degenerate DR-1 site, and physically blocks AP-1 and transcriptional activation by IL-1β. (From Eck *et al.*, 2009.)

found in MMP promoters, raising the question as to how glucocorticoids can be so effective in repressing MMP gene expression. These receptors can regulate transcription by interfering with the ability of AP-1 proteins to bind to AP-1 sites and/or the core of proteins that bind to the TATA box, the master regulator of transcription. It is thought that these activated glucocorticoids receptors complex with the AP-1 proteins that are already bound to the DNA of MMP promoters, inducing a conformational change in AP-1 proteins that represses transcription. These findings emphasize the critical role of AP-1 sites in controlling MMP gene expression and suggest that the ubiquitous presence of these sites in MMP promoters (and many other genes as well) prevents the steroids from specifically targeting MMP genes.

Retinoids also suppress MMP transcription through several mechanisms. First, like the glucorticoids, they interfere with the binding of AP-1 proteins to the all-important proximal AP-1 site, although upstream sequences may contribute.[4] Second, RAR may bind directly to c-Jun and sequester it, thus preventing its binding to DNA. Third, retinoids downregulate mRNAs for c-fos and c-jun proteins, which are the principal transcription factors that drive the AP-1 site.

More recently, it has been reported that the proximal AP-1 site in the MMP-1 and MMP-13 promoters overlaps with a composite DR-1 element in which a nuclear hormone receptor and AP-1 bind to the DNA in a mutually exclusive manner to reduce MMP-1 expression.[8] Other DR-1 elements may be located upstream in several other MMP promoters, suggesting cooperativity among *cis*-acting sequences as another mechanism by which retinoids may downregulate MMP transcription.

Summary. The transcriptional regulation of MMPs is complex, requiring the integration of multiple signaling pathways and the activation of a myriad of transcription factors that bind to a cohort of *cis*-acting elements within any given MMP promoter. As a result, different mechanisms may control transcription of different MMPs in different cells and tissues. These mechanisms probably depend on the assortment of transcription factors found within a particular cell type, the nature of the inducer or repressor that binds to its receptor, and the subsequent initiation of signal transduction pathways and activation of transcription factors. The "mixing and matching" of these mechanisms provides a wide range of potential interactions among transcription factors and *cis*-acting sequences, which may explain how tissue specificity is achieved among the different members of the MMP family.

3.7 Posttranscriptional Mechanisms

1. *mRNA stability.* Although MMPs are mainly regulated at the transcriptional level, stabilizing mRNAs to increase their half-life

once they are in the cytoplasm is another important mechanism for regulating the expression of many genes, including MMPs.[4,7–9] The mRNAs of several MMPs, including MMP-1, are inherently unstable, perhaps because MMP-1 protein can have such deleterious effects on the ECM. Consequently, a short mRNA half-life represents another mechanism of regulating gene expression. Therefore, under conditions of normal homeostasis, only low levels of steady-state mRNA are detected, despite substantial rates of basal transcription.[4] As we know, cytokines and growth factors, such as IL-1β and EGF, are potent inducers of MMP transcription, but importantly, they also increase mRNA stability.[4] The net result is higher levels of steady-state mRNA and greater potential for facilitating matrix remodeling/destruction under certain conditions.

MMP transcripts are stabilized through specific sequences in their 3′-UTRs. These sequences are potential targets of different UTR-binding proteins that can either stabilize or destabilize these mRNAs (Fig. 3.8).[8,9,30] For example, rabbit and human MMP-1 mRNAs contain three copies of the motif, 5′-AUUUA-3′, "AU-rich elements (AREs)", which bind proteins that increase mRNA stability, and mutation of these motifs abolishes the increased half-life seen with cytokine and growth factor treatment.[4] In another example, rat MMP-9 mRNA contains several copies of these AU-rich elements within its 3′-UTR, and MMP-9 mRNA that is

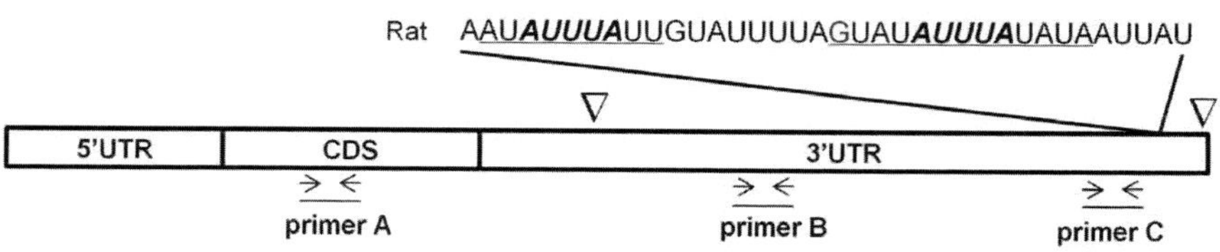

Figure 3.8. Example of AUUA sequences in 3′-UTR of mRNA. A highly conserved cluster of ARE in the 3′-UTR, with a consensus core sequence of AUUUA flanked by symmetric A or U (underlined). Triangles indicate alternative polyadenylation sites in the primary transcript. (From Allen *et al.*, 2013.)

transcriptionally increased following stimulation with IL-1β is further increased due to mRNA stabilization.[9] In contrast, a decrease in MMP-9 mRNA can be seen if there is a reduction in the levels of a 3′-UTR binding protein.[9] Thus, in concert with the transcriptional induction of MMPs, enhancing mRNA stability provides an additional means for regulating (or perhaps fine-tuning) gene expression.[4] On the other hand, on some occasions, mRNAs can be destabilized.

2. *Micro-RNAs.* Recent studies reveal an additional mechanism for regulating mRNAs at a posttranscriptional level: miRNAs, which silence mRNAs. MiRNAs are a large family of noncoding single-stranded RNAs of ~22 nucleotides (nt) in length (range 19–25 nt).[31,32] The genes encoding miRNAs are usually found in regions between genes or embedded in the introns of functional genes. They are transcribed by RNA polymerase II into hairpin structures called "primary microRNAs (pri-miRNAs)". These are processed to pre-miRNAs in the nucleus by the RNAse III enzyme, Drosha, and the double-stranded RNA-binding proteins are exported to the cytoplasm (Fig. 3.9; www.sigmaaldrich.com). There they are cleaved by yet another RNase III type enzyme, Dicer. Dicer generates a double-stranded RNA of ~22 nt; one strand becomes a mature miRNA, which is assembled into an RNA-induced silencing complex (RISC), while the other strand is degraded. The RISC complex interacts with the "Argonuate" proteins, which together "silence" the target mRNAs, most often by binding to target sites for miRNAs, which are usually located in the 3′-UTR of the target mRNA. The degree of silencing of the mRNA depends on the amount of complementarity. With total complementarity of aligned miRNA/mRNA pairs, the mRNA is degraded by endonucleolytic cleavage. However, in most cases, complementarity is imperfect, and rather than the mRNA being degraded, its translation into protein is suppressed. Thus, when documenting effects of miRNAs, the absence of translated protein, rather than the absence of mRNA is the appropriate focus.

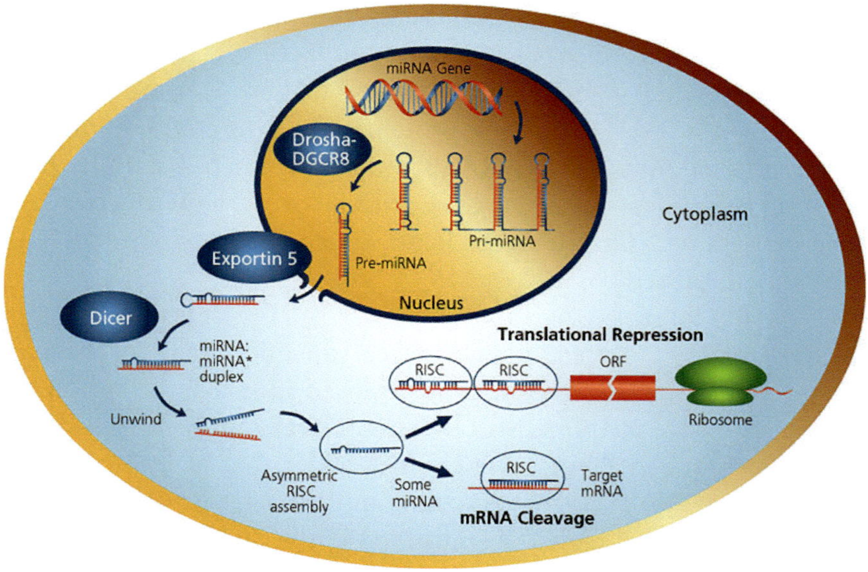

Figure 3.9. Scheme of micro-RNA synthesis and function. (From www.sigma aldrich.com.)

Identifying miRNAs and their target genes is a rapidly growing area of investigation and it has been suggested that 30% of the human genome may be regulated by this mechanism.[8,9,31,32] Some of the effects of miRNAs are indirect. For example, miRNAs may act as a tumor suppressor gene by suppressing expression of Ras.[8] Conversely, when miRNA-21 is overexpressed, it functions as an oncogene by downregulating expression of pro-apoptotic genes.

At this point, reports of miRNAs modulating activities of MMPs are largely anecdotal, with no coalescing of patterns of miRNAs regulating particular MMPs. Most likely, this is a function of the relatively early stages in our knowledge of these regulatory molecules. Nonetheless, it is useful to provide a few examples of effects of miRNAs on MMPs. In some cases, MMPs expression is modulated indirectly through miRNAs, which target genes that are part of different signaling pathways responsible for MMP activation.[8,31,32] However, miRNAs can directly target MMPs. Several reports document a role for miRNAs in modulating MMPs in rheumatoid arthritis and osteoarthritis.[31] Increased miR-203 is associated with increased MMP-1, an effect mediated by NF-κB signaling in fibroblasts from rheumatoid tissues.[31] In addition, miR-27b downregulates MMP-13 in both normal and osteoarthritic chondrocytes, and miR-146a is involved in cartilage degradation in osteoarthritis.[31] Similarly, in osteoarthritis, MMP-13 protein levels are affected by miR-9 reducing TNF-α.[8]

In cancers, MMP-2 is a target of miR-29b in prostate cancer cells,[32] and miR-206 downregulates MMP-2 and MMP-9 in breast cancer cells by suppressing tumor cell invasion and migration by remodeling the actin cytoskeleton.[32] More than 50 miRNAs have been implicated in modulating the expression of several MMPs in cancers,[32] with reports on miRNAs and MMP-2 and MMP-9 in cancer cells outweighing reports on other MMPs.[32] While it may be the case that these two MMPs are preferentially regulated by miRNAs, it is also possible that this is simply a reflection of current investigations and that future studies will implicate additional MMPs since bioinformatic analyses have predicted potential miRNA-binding sites in the 3′-UTR regions of several MMPs.[8]

Based on the growing relevance of miRNAs in the regulation of biological and pathological processes, it is virtually certain that further work will unveil the identity and roles of additional miRNAs in targeting specific MMPs and thus, in contributing to their functions in a variety of contexts.[8,31,32]

3.8 Epigenetic Regulation

"Epigenetics" describes the study of stable, long-term changes in the transcriptional potential of cells. In contrast to "genetics", which is based on changes to the DNA sequence (i.e., the genotype), epigenetics causes changes in the expression of genes that change the phenotype of cells, that is, how they behave. The term is derived from the Greek term *epi-* (Greek: *επί-* over, outside of, around). Thus, epigenetics is outside/around genetics. Acetylation of histone proteins and methylation of CpG islands in DNA are epigenetic regulators of transcriptional initiation, and thus, are powerful adjuncts in controlling gene expression.[7–9]

1. *Acetylation* is a posttranslational and covalent modification of proteins that are associated with transcription. Acetylation is controlled by enzymes, which modify histones and complexes of protein that remodel chromatin.[7–9] The sites on histones where modifications occur are docking places for chromatin-modifying complexes (such as switch/sucrose nonfermentable [SWI/SNF]) (Fig. 3.10). These complexes twist the DNA, and as a result, the position of nucleosomes (the units of chromatin) is shifted to either a permissive (favoring transcription) or nonpermissive (unfavorable for transcription) conformation. Therefore,

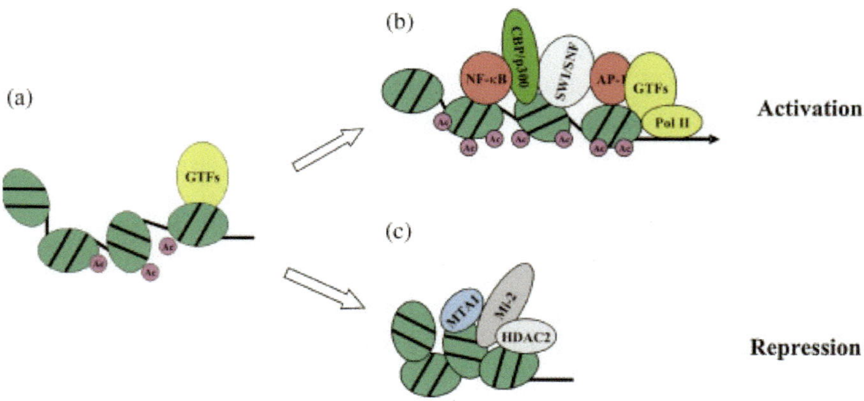

Figure 3.10. Epigenetic mechanisms contribute to both activation and repression of MMP-9 gene expression. The MMP promoter is preassembled with the general transcription factors (GTFs) such as TATA-binding protein (TBP), transcription factor (TF) IIA, IIB, IID, IIE, IIF, IIH, and low levels of polymerase II (Pol II) to maintain the basal expression of the gene (a). The conversion of chromatin structure by recruitment of either transactivator/coactivator complexes such as NF-κB, AP-1, CBP/p300, and SWI/SNF (b) or corepressor complexes such as the nucleosome remodeling and deacetylase (NuRD) complex, which contains MTA1, Mi-2, and HDACs (c). Alters access of the MMP-9 promoter to Pol II, thereby activating or repressing expression of this gelatinase. Ac represents acetylated lysine residues in histone H3 or H4. (From Yan and Boyd, 2006.)

epigenetic modulation of transcription requires the recruitment and cooperation between transcription factors that will bind to sequences in the DNA and complexes of proteins that remodel/reshape chromatin.

Histones carry a positive charge, while the charge on DNA is negative. The covalent addition of an acetyl group to the histones reduces their charge, thereby reducing the strength of the histone–DNA interaction and allowing increased access of transcription factors to the DNA. Thus, acetylation is usually associated with gene activation. The core histones (H3, H4, and H2B) are acetylated by a family of enzymes called histone acetyl transferases (HATs). However, acetylation is a reversible process, and acetyl groups are removed by histone deacetylases (HDACs), and once deacetylation has occurred, transcription is suppressed.[7–9]

The effects of the acetylation and chromatin remodeling in controlling MMP expression have been reported.[7–9] The induction of MMP-1 and MMP-13 by IL-1α and oncostatin M is almost completely abolished by inhibitors that block the removal of acetyl groups (HDAC inhibitors). This indicates that acetylation has a critical role in regulating transcription of these MMPs in response to inflammatory stimuli.[8,9] In another study, modification of acetylated histone H4 (AcH4) was induced by IL-1β throughout the entire region of the MMP-13 promoter, including a distal AP-1 site ~20 kb upstream of the transcription start site,[9,14] emphasizing the importance of chromosome looping in contributing to gene regulation.

However, acetylation, alone, may not be sufficient for inducing MMP-1 expression. This suggests that initiation of MMP-1 transcription first requires activation of certain transcription factors, such as c-Jun, c-Fos, followed by their binding to the MMP-1 promoter and finally, by the recruitment of histone acetyltransferases. All of these changes result in a state of the DNA that is permissive for initiating transcription.[8,9] Indeed, the sequential assembly of transcription complexes on the MMP-9 promoter has been demonstrated. This process depends on the activation of the MEK-1/ERK and NF-κB signaling pathways, followed by the recruitment of transcription factors, chromatin-remodeling complexes, and coactivators. All are assembled at the MMP-9 promoter in a stepwise and coordinated manner. Therefore, the many mechanisms that regulate MMP expression require multiple complex events that must operate in a coordinated manner.

2. *Methylation of DNA* in eukaryotic cells happens only on cytosine bases that are located within CpG nucleotide islands in the promoters of genes. Methylation is an effective means of repressing transcription, and is often associated with inactive chromatin and an inhibition of gene expression (Fig. 3.11).[7–9] DNA methylation occurs through the action of DNA methyltransferases (DNA MTase), a family of enzymes that catalyzes the transfer of a methyl group to DNA. Methylation of DNA interferes with the ability of transcriptional activators to bind to promoter DNA.

Several effects of methylation (or lack thereof) on the expression of several MMPs have been documented. Not surprisingly, an increase in promoter methylation of MMP-9 is linked to a decrease in gene expression in lymphoma cells.[7,9] In contrast, a reduction in methylation

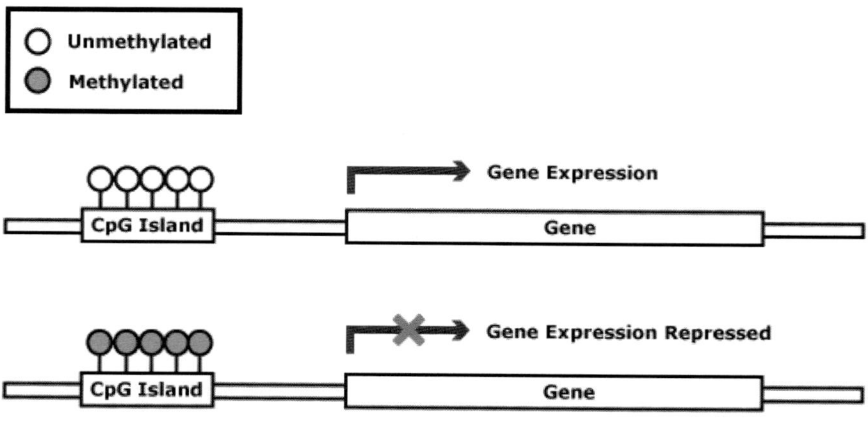

Figure 3.11. DNA methylation. (From UCSF School of Educational Technology, 2007.)

(hypo-methylation) of MMP-2 increased gene expression with a concomitant increase in tumorigenesis and tumor cell invasiveness. Similarly, colon cancer cells that are defective in two DNA methyltransferases (Dmt-1 and Dmt3b) have an increase in MMP-3 expression. However, increasing MMP-3 gene expression by hypomethylation was cell-specific, since methylase inhibitors failed to increase MMP-3 in a lymphoma cell line. As a final example, chondrocytes from patients with osteoarthritis have higher levels of MMP-3, MMP-9, and MMP-13, which correlate with a decrease in methylation of the CpG islands in these promoters.[8]

In conclusion, acetylation of histones, methylation of DNA, and remodeling of chromatin are not mutually exclusive.[7] Rather, they cooperate in conjunction with the activation of signal transduction pathways and transcription factors to maximally induce (or repress) gene expression. Future studies will reveal additional molecular details as to how all of these mechanisms align to regulate the expression of genes, including the MMPs, in a larger context that is associated with other sequences within the gene and/or with other genes within the cell and organism as a whole.

3.9 Summary

Over past decades, much information on the transcriptional regulation of MMP gene expression has been consolidated; the promoters of most MMP family members have been well characterized. AP-1 sites are commonly involved with other *cis*-acting sequences, such as ETS, RUNX, and NF-κB, contributing to transcriptional regulation in an MMP- and tissue-specific manner. At the level of gene expression, many hormones, cytokines, and growth factors are potent inducers of MMP expression. Tissue specificity of expression of individual MMPs is mainly achieved by the combination of different transcriptional control mechanisms. In addition, several SNPs have been discovered in MMP promoters and these have the ability to augment or suppress MMP transcription.

The integration of multiple signaling pathways, coupled with the cooperation between several *cis*-regulatory elements of the MMP promoters, facilitate the strict spatiotemporal control of MMP transcriptional activity. Additionally, epigenetic mechanisms, such as DNA methylation or histone acetylation, contribute to MMP regulation. Further, posttranscriptional mechanisms regulating mRNA stability, protein translational efficiency, and microRNA-based mechanisms have been described. MMP polymorphisms, which may contribute to disease pathologies, have been identified. Nonetheless, the fundamental mechanisms that initiate transcription of a gene remain at the center of investigation, for without this initial response, the other contributing mechanisms of gene regulation become moot. Whether we are discussing signal transduction pathways, *cis*-acting sequences within a promoter, or SNPs, these elements are not functioning as independent events, but rather in a larger context that is associated with other sequences within the gene and/or with other genes within the cell and organism as a whole.

References

1. Boyle, J (2008). Molecular biology of the cell. In *Biochemistry and Molecular Biology Education*, B Alberts, A Johnson, J Lewis, M Raff, K Roberts and P Walter (eds.), 5th Ed., Vol. 36. Garland Press, New York, New York.
2. Gaffney, J, I Solomonov, E Zehorai and I Sagi (2015). Multilevel regulation of matrix metalloproteinases in tissue homeostasis indicates their molecular specificity in vivo. *Matrix Biology: Journal of the Internation Society for Matrix Biology*, 44–46, 191–199. doi:10.1016/j.matbio.2015.01.012.
3. Angel, P and M Karin (1991). The role of Jun, Fos and the AP-1 complex in cell-proliferation and transformation. *Biochimica et Biophysica Acta — Review Cancer*, 1072, 129–157.
4. Vincenti, MP *et al.* (1996). Regulating expression of the gene for matrix metalloproteinase-1 (collagenase): mechanisms that control enzyme activity, transcription and mRNA stability. *Critical Reviews Eucaryotic Gene Expression*, 6(4), 391–411.
5. Benbow, U and CE Brinckerhoff (1997). The AP-1 site and MMP gene regulation: what is all the fuss about? *Matrix Biology: Journal of the Internation Society for Matrix Biology*, 15, 519–526.

6. Brinckerhoff, CE and LM Matrisian (2002). Matrix metalloproteinases: a tail of a frog that became a prince. *Nature Reviews Molecular Cell Biology*, 3, 207–214.
7. Yan, C and DD Boyd (2006). Regulation of matrix metalloproteinase gene expression. *Journal of Cellular Physiology*, 211, 19–26. doi:10.1002/JCP.20948.
8. Clark, IM, TE Swingler, CL Sampieri and DR Edwards (2008). The regulation of matrix metalloproteinases and their inhibitors. *International Journal of Biochemistry and Cell Biology*, 40, 1362–1378.
9. Fanjul-Fernández, M, AR Folgueras, S Cabrera and C López-Otín (2010). Matrix metalloproteinases: evolution, gene regulation and functional analysis in mouse models. *Biochimica et Biophysica Acta*, 1803, 3–19.
10. Angel, P *et al.* (1987). 12-0-Tetradecanoyl-phorbol-13-acetate induction of the human located in the 5′-flanking region. *Molecular and Cell Biology*, 7, 2256–2266.
11. Angel, P *et al.* (1987). Phorbol ester-inducible genes contain a common cis element recognized by a TPA-modulated mans-acting factor. *Cell*, 49, 729–739.
12. Angel, P and M Karin (1991). The role of Jun, Fos and the AP-1 complex in cell-proliferation and transformation. *Biochimica et Biophysica Acta*, 1072, 129–157.
13. Gutman, A and B Wasylyk (1990). The collagenase gene promoter contains a TPA and oncogene-responsive unit encompassing the PEA3 and AP-1 binding sites. *EMBO Journal*, 9, 2241–2246.
14. Schmucker, AC, JB Wright, MD Cole and CE Brinckerhoff (2012). Distal interleukin-1β (IL-1β) response element of human matrix metalloproteinase-13 (MMP-13) binds activator protein 1 (AP-1) transcription factors and regulates gene expression. *Journal of Biological Chemistry*, 287, 1189–1197.
15. Johnson, JL *et al.* (2012). Regulation of matrix metalloproteinase genes by E2F transcription factors: Rb-Raf-1 interaction as a novel target for metastatic disease. *Cancer Research*, 72, 516–526.
16. Foley, CJ and A Kuliopulos (2014). Mouse matrix metalloprotease-1a (Mmp1a) gives new insight into MMP function. *Journal of Cellular Physiology*, 229, 1875–1880. doi:10.1002/jcp.24650.
17. Rodbell, M (1995). Signal transduction: evolution of an idea. *Bioscience Report*, 15, 117–133.

18. Gilman, AG (1995). G proteins and regulation of adenylate cyclase. *Bioscience Reports*, 15, 65–97.
19. Vincenti, MP and CE Brinckerhoff (2007). Signal transduction and cell-type specific regulation of matrix metalloproteinase gene expression: can MMPs be good for you? *Journal of Cellular Physiology*, 213, 355–364. doi:10.1002/JCP.
20. Rodbell M (1995). Nobel Lecture. Signal transduction: evolution of an idea. *Bioscience Reports*, 15, 117–133..
21. Whipple, CA and CE Brinckerhoff (2014). BRAF(V600E) melanoma cells secrete factors that activate stromal fibroblasts and enhance tumorigenicity. *British Journal of Cancer*, 111, 1625–1633.
22. Blackburn, JS, CH Rhodes, CI Coon and CE Brinckerhoff (2007). RNA interference inhibition of matrix metalloproteinase-1 prevents melanoma metastasis by reducing tumor collagenase activity and angiogenesis. *Cancer Research*, 67, 10849–10858.
23. Blackburn, JS, I Liu, CI Coon and CE Brinckerhoff (2009). A matrix metalloproteinase-1/protease activated receptor-1 signaling axis promotes melanoma invasion and metastasis. *Oncogene*, 28, 4237–4248.
24. Leivonen, SK and VM Kähäri (2007). Transforming growth factor-beta signaling in cancer invasion and metastasis. *International Journal of Cancer*, 121, 2119–2124.
25. Jenkins, MH, W Croteau, DW Mullins and CE Brinckerhoff (2015). The BRAF(V600E) inhibitor, PLX4032, increases type I collagen synthesis in melanoma cells. *Matrix Biology: Journal of the Internation Society for Matrix Biology*, 1–12. doi:10.1016/j.matbio.2015.05.007.
26. Schroen, DJ and CE Brinckerhoff (1996). Nuclear hormone receptors inhibit matrix metalloproteinase gene expression through diverse mechanisms. *Gene Expression*, 6, 197–207.
27. Rutter, JL *et al.* (1998). A single nucleotide polymorphism in the matrix metalloproteinase-1 promoter creates an Ets binding site and augments transcription. *Cancer Research*, 8, 5321–5325.
28. Brinckerhoff, CE, JL Rutter and U Benbow (2000). Interstitial collagenases as markers of tumor progression. *Clinical Cancer Research*, 6, 4823–4830.
29. Brinckerhoff, CE and MB Sporn (2003). Retinoids and rexinoids for the 21st century: a brave new world for arthritis. *Journal of Rheumatology*, 30, 4–7.
30. Allen, M *et al.* (2013). HuD promotes BDNF expression in brain neurons via selective stabilization of the BDNF long 3′-UTR mRNA. *PLoS One*, 8, e55718.

31. Jeyapalan, Z, TN Wight and BB Yang (2013). miRNAs regulate expression and function of extracellular matrix molecules. *Matrix Biology: Journal of the Internation Society for Matrix Biology*, 32, 74–85.
32. Abba, M, N Patil and H Allgayer (2014). MicroRNAs in the regulation of MMPs and metastasis. *Cancers (Basel)*, 6, 625–645.
33. Eck SM, JS Blackburn, AC Schmucker, PS Burrage and CE Brinckerhoff (2009). Matrix metalloproteinase and G protein coupled receptors: co-conspirators in the pathogenesis of autoimmune disease and cancer. *Journal of Autoimmunity*, 33, 214–221.

4. "Matrix Metalloproteinases: They're Not Just for Matrix Anymore!"

Constance E. Brinckerhoff

4.1 Introduction

As the number of matrix metalloproteinase (MMP) family members continued to grow, interest in their functions exploded. Initially, mammalian MMPs were identified as potent modulators of the extracellular matrix (ECM) in tissue remodeling associated with normal developmental processes and with disease.[1] Subsequently, however, it became clear that matrix proteins are not the only physiological, or even pathological, substrates. They are, in fact, "not just for matrix anymore".[1] MMPs are now known to control many other aspects of cellular responses to their environment: modifying the activity of membrane proteins implicated in cell–cell interactions, activating cytokines and growth factors, and their receptors,[1–3] and more recently, displaying unexpected pericellular and even intracellular functions.[2–5] Thus, the current concept of MMP functions has progressed from the original, historical view, as mediators of matrix degradation to encompass the realization that MMPs have many diverse functions, which place them at the "cutting edge" of biology. Consequently, the sophisticated and complex regulatory mechanisms governing their expression and enzymatic activities, discussed in previous chapters, take on even greater significance as they are applied to a seemingly ever-growing list of functions.

The fact that MMPs can degrade all components of the ECM gives them a critical position in modeling and remodeling connective tissues in health and disease. That, alone, would assure them a significant place in biology. However, in the process of degrading matrix, they also generate new proteins with new functions. For example, cleavage of collagen XVIII by MMP-7 (or the cathepsins) generates endostatin, an antiangiogenic compound.[6,7] In addition, MMPs can activate and liberate growth factors and hormones embedded in our connective tissues, thereby increasing bioavailability and influencing the cellular microenvironment. By acting on these nonmatrix substrates, the repertoire of MMPs is expanded hugely, as is their impact on biological systems.

4.2 Recognizing New Substrates

Traditionally, MMPs were seen as "bulldozers",[1] ruthlessly destroying connective tissues as part of the pathological processes of disease progression. However, as the number of MMP family members grew and the process of identifying ECM substrates continued, it became clear that viewing MMPs only as enzymes that can degrade ECM molecules is naïve and simplistic.[1,2] Instead, it has been suggested that, MMPs should be regarded as proteinases that cleave specific sequences within proteins, and consequently target matrix and nonmatrix substrates (Fig. 4.1).[2] Understanding the molecular basis by which MMPs recognize these divergent substrates and documenting the specific sequences that are proteolytically attacked remain important and unfinished areas of investigation.

The ever-broadening portfolio of MMP substrates has led to the suggestion that the total amount of proteolytic activity in a biological system depends on interactions between the different families of proteases and protease inhibitors, all of which are linked in a protease web. MMPs are central components of this web, which is increasingly recognized as a complex interaction among the proteolytic players.[2] The concept of proteomics and degradomics has emerged as the characterization of all proteases, inhibitors, and protease substrates by genomic and proteomic techniques. The concept

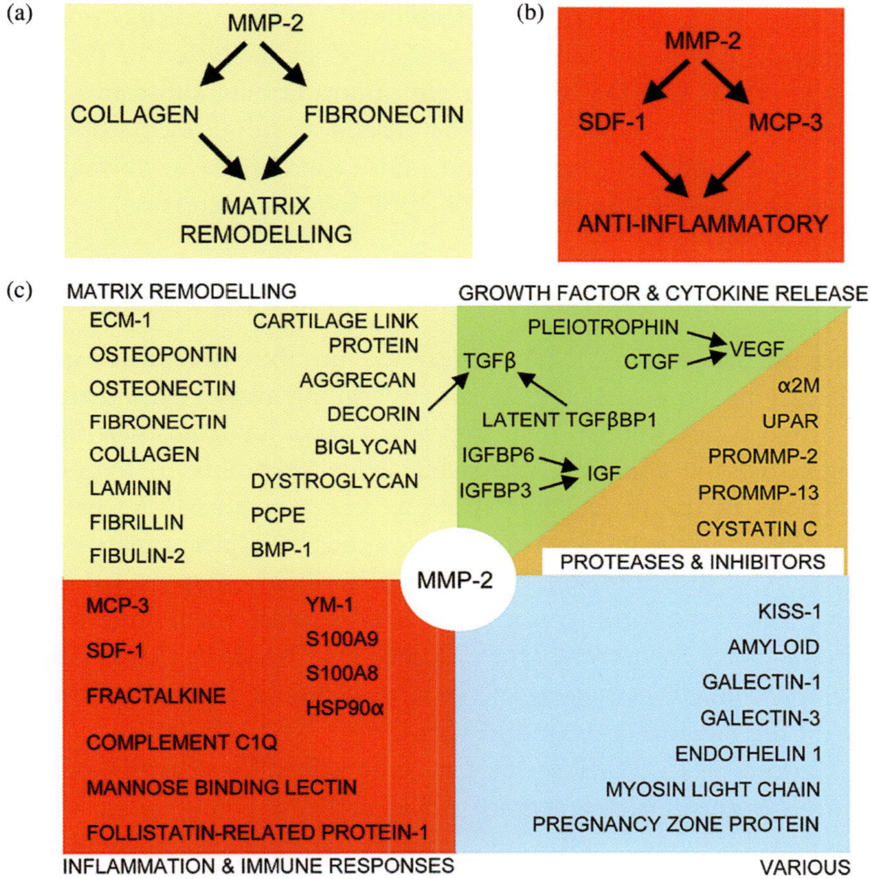

Figure 4.1. Substrates define protease function. Substrates such as collagen and fibronectin imply a role for MMP-2 in ECM remodeling (a), whereas the substrates stromal-derived factor-1 (SDF-1/CXCL12) and monocyte chemoattractant protein-3 (MCP-3/CCL7) advocate an anti-inflammatory role (b). In fact, MMP-2 can cleave a large number of substrates [MEROPS (http://merops.sanger.ac.uk/)] and therefore is multifunctional (c). (From Butler and Overall, 2009.)

has gained credibility as it has "exponentially expanded the known substrate repertoire of the MMPs, even to include intracellular proteins with newly recognized extracellular functions. Thus, the dogma that MMPs are dowdy degraders of ECM has been resolutely overturned, and the metamorphosis of MMPs into modulators of multiple signaling pathways has been facilitated".[2] The identification of numerous nonmatrix substrates and the documentation of the biological effects of MMPs in cleaving nontraditional substrates indicate that these enzymes should be viewed as subtle mediators that they have enormously varied and multifaceted roles in normal cell behavior, cell–cell communication, and disease progression.

4.3 Pericellular Activities

Since the majority of MMPs are secreted into the extracellular space, it is easy to think of MMPs as simply "sailing away" from their cell of origin to accomplish their appointed tasks far from their site of origin. To the contrary, increasing evidence points to important functions in the pericellular space as the secreted MMPs become sequestered in the microenvironment, either by adjacent matrix proteins and/or in a complex with other pericellular components. Typically, one associates pericellular MMPs with membrane-bound MMPs. Since membrane-type 1-MMP (MT1-MMP) (MMP-14) is the best studied, perhaps it can be considered as the "senior" member of the membrane-bound MMPs.

Indeed, MT1-MMP seems especially suited for pericellular proteolysis because it can degrade adjacent ECM components, transmembrane proteins, and soluble factors.[5] Further, MT1-MMP may also function intracellularly by helping to process intracellular substrates, including, but not limited to, pro-alpha v integrin and focal adhesion kinase (Fig. 4.2).[2,5] Although the mechanism(s) of proteolytic activity inside the cell is not clear, it is clear that localized compartmentalization of MT1-MMP (either inside or outside the cell) is an essential component of its diverse proteolytic functions. Further, the presence of MT1-MMP in caveolae/lipid rafts of the cell membrane may help to direct both its location and its activities. For example, forcing MT1-MMP into lipid rafts by deleting its cytosolic tail limited its ability to access and process E-cadherin, and this limited its ability to influence tumor cell locomotion and growth.[5] Usually, however, if MT1-MMP is found intracellularly, it is largely sequestered in membrane vesicles and may represent either newly

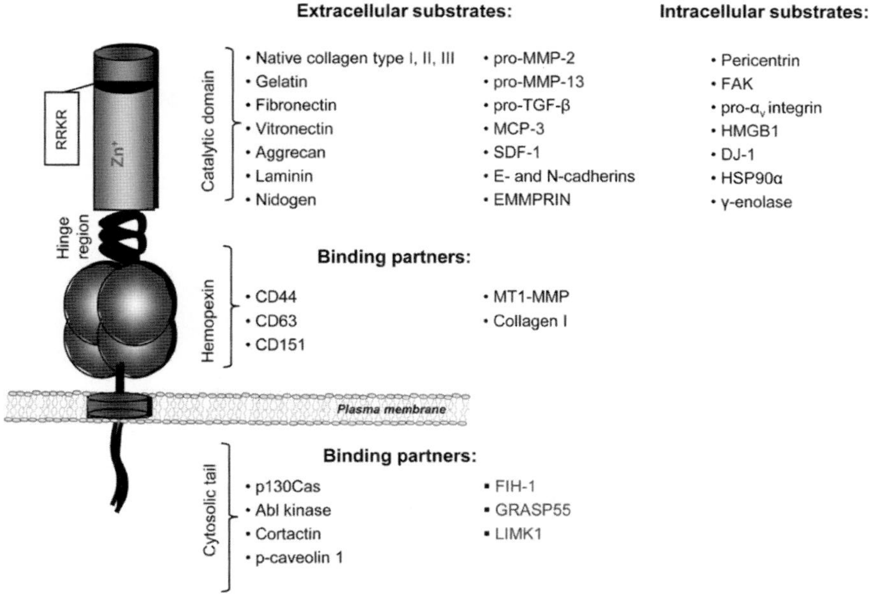

Figure 4.2. Extracellular and intracellular substrates and partners of MT1-MMP. The domain structure of MT1-MMP is shown together with lists of selected extracellular and intracellular substrates and binding partners associated with each domain. Intracellular binding partners that interact with MT1-MMP at the Golgi apparatus are indicated in blue text. (From Koziol *et al.*, 2012.)

synthesized protein or surface enzyme that has been recycled through endocytosis.

In addition, since MT1-MMP is concentrated at lamellipodia, which is the migration front of cells, it is associated with localized degradation of the ECM (Fig. 4.3).[5,8] The short 20 amino acid cytoplasmic domain of MT1-MMP is important in linking MT1-MMP to efficient cell migration and invasion; deleting this domain reduces MT1-MMP-mediated cell migration, but does not affect its cell-surface proteolytic activity. However, other studies in which MT1-MMP without its cytoplasmic tail was overexpressed still showed cell migration. These results remain controversial because it is possible that ectopic overexpression overloaded the secretory pathway with an excess of enzyme, thereby overriding the normal regulatory

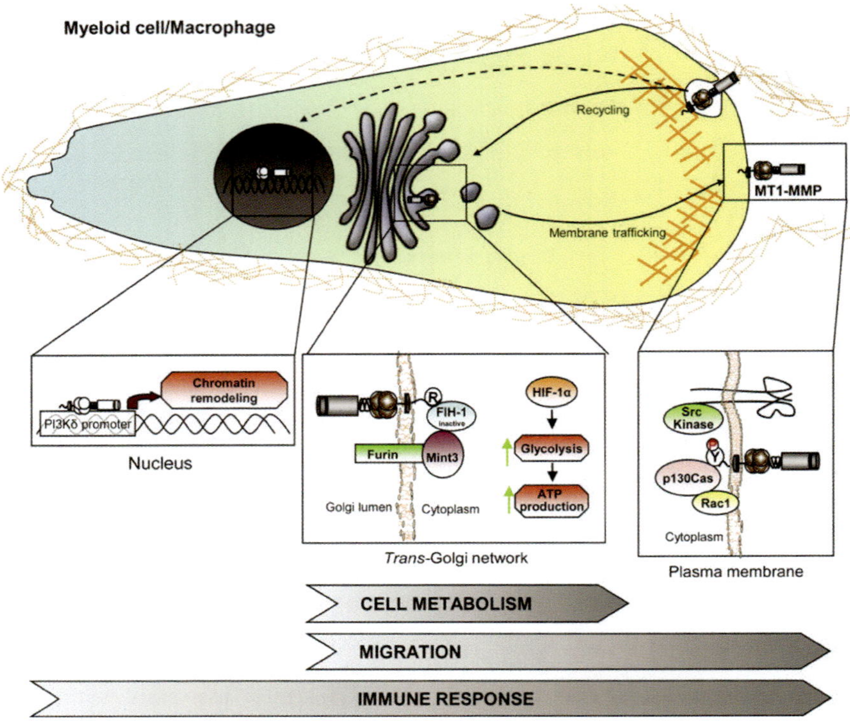

Figure 4.3. Site-specific functions of MT1-MMP in macrophages. MT1-MMP is involved in macrophage migration and the immune response in a catalytic-independent manner through at least three site-specific pathways linked to distinct MT1-MMP traffic routes. At the plasma membrane, MT1-MMP can modulate Rac1 activity through binding to p130Cas, thereby regulating migration. The formation of this complex can be regulated by tyrosine phosphorylation of MT1-MMP cytosolic Y573 by Src kinase. At the Golgi apparatus, MT1-MMP, through the Arg576 in its cytoplasmic tail, binds factor-inhibiting HIF-1 (FIH-1), keeping it close to Mint-3. This inactivation of FIH-1 stabilizes hypoxia inducible factor-1 (HIF-1), resulting in the activation of glycolytic adenosine triphosphate (ATP) production needed for macrophage migration and the immune responses. Upon translocation to the nucleus through as-yet undefined mechanisms, MT1-MMP induces phosphoinositide -3 kinase (PI3K) expression and regulates the transcription of genes critical for the inflammatory response. (From Koziol *et al.*, 2012.)

mechanisms. Nonetheless, the transmembrane domain is essential because secreted collagenases, such as MMP-1, or MT1-MMP lacking this domain could not mediate cell invasion through collagen.

MT1-MMP, therefore, serves as a model for understanding pericellular and intracellular functions of MMPs because of its particular structure and function as it straddles the cell membrane. However, we are learning that even the secreted MMPs can function pericellularly with novel effects on cellular behavior in a microenvironment where they regulate remodeling events and direct cell migration and interactions with other cells. MMPs secreted into the extracellular space may be recruited back to the local cell environment by interactions with cell surface proteins and the pericellular matrix (Table 4.1).[8] Under these conditions, MMPs are not functioning singularly, but rather seem to be part of a cluster of proteins that act together to accomplish a particular task.[8,9]

Table 4.1. Cell surface-binding partners of MMPs. Some representative examples of the associations of MMPs with cell surface molecules and the domain involved in interactions are shown. It is possible that such interactions confer site specificity for MMP action and could form the basis for therapeutic targeting outside the catalytic cleft.

MMP	Domain	Cell surface-binding
MMP-1	Linker and Hemopexin	$\alpha 2\beta 1$ integrin I-domain
MMP-2	Hemopexin	MTI-MMP:TIMP-2 complex
MMP-2	Hemopexin	Chondroitin 4-sulfate (cell membrane)
MMP-2	Catalytic	$\beta 2$ integrin
MMP-2	FN domain	LRP-1
MMP-2	?	Bone sialoprotein (cell surface)
MMP-3	Hemopexin	Collagen I
MMP-7	?	CD44, syndecans, glypicans (heparin sulfate)
MMP-7	Propeptide	CD151
MMP-7	Catalytic	Cholestrol sulfate (cell membrane)
MMP-8	Pro and active forms	Neutrophil surface

(Continued)

Table 4.1. (*Continued*)

MMP	Domain	Cell surface-binding
MMP-9	Hemopexin	CD44
MMP-9	Hemopexin	Ku dimer (Ku80)
MMP-9	Hemopexin	β1 integrin
MMP-9	Catalytic	β2 integrin
MMP-9	Hemopexin	β5 integrin I-EGF-like domains 2 and 3
MMP-9	Not known	Type VI collagen α2 chain (cell surface)
MMP-9	Linker, hemopexin	LRP-1 and -2
MMP-13	Not known	Receptor/LRP-1
MMP-14 (MTI-MMP)	Hemopexin	CD44H stem
MMP-14	Not known? Indirect	Claudin
MMP-14	Hemopexin	CD63
MMP-14	Hemopexin	MMP-14
MMP-14	?	CD151
MMP-15	Hemopexin	CD44H
MMP-16	Hemopexin	CD44H
MMP-16	Catalytic	Chondroitin-4-sulfate (cell surface)
MMP-24	Hemopexin	CD44H
MMP-25	Hemopexin	CD44H

Source: Murphy and Nagase (2011).

Exactly how secreted MMPs behave in pericellular space may depend on their interactions with particular matrix proteins, where interactions between MMPs and matrix proteins influence cell behavior and matrix remodeling. In a fascinating example of specialization, studies of wound healing in keratinocyte monolayers have shown that the collagenolytic activity of MMP-1 is necessary for migration of human keratinocytes on type I collagen (Fig. 4.4).[10] With wounding, the integrin α2β1 in the keratinocytes makes contact with collagen fibrils, inducing the cells to produce proMMP-1. Then, the hemopexin domain of both proMMP-1 and MMP-1 binds to the α2β1 integrin through the α2 integrin subunit. The α2β1 integrin binds native collagen I with high affinity, resulting in clustering of this integrin at

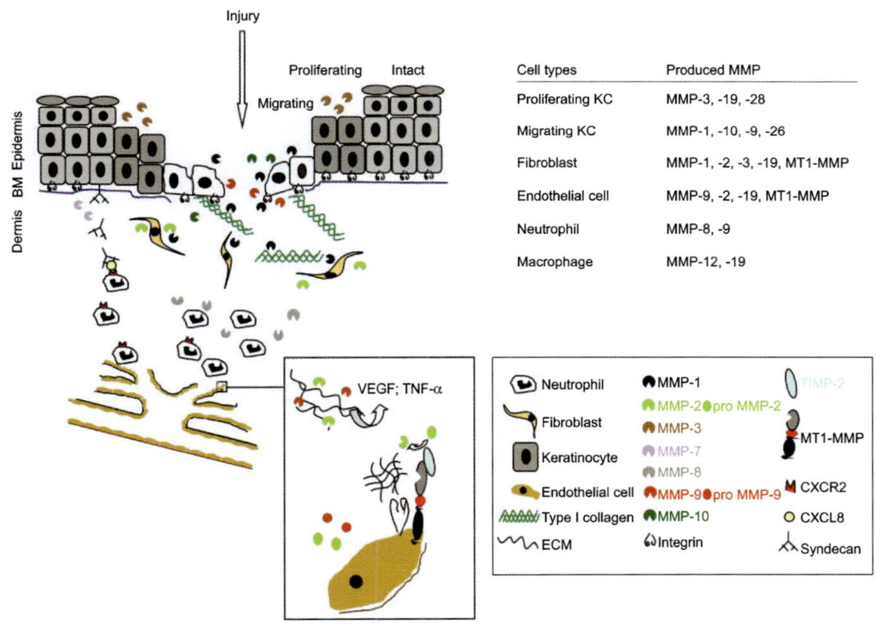

Figure 4.4. Distinct functions of MMPs contribute to wound healing and angiogenesis. In healthy skin, basal expression of MMPs is low, but following injury, interaction of many secreted and membrane-bound MMPs contribute either directly or indirectly to wound healing. For example, MMP-1 strongly influences keratinocyte migration at the wound edges, while MMP-7 regulates neutrophil recruitment (as shown in the legend on the lower right site). The wound-induced neoangiogenesis, a result of migration and proliferation of ECs, is driven by VEGF and tumor necrosis factor (TNF)-α release through the gelatinases MMP-2 and MMP-9 (as shown in the box with a blow up of a migrating EC. BM, basement membrane; EC, endothelial cell; ECM, extracellular matrix; KC, keratinocyte; TIMP, tissue inhibitor of matrix metalloproteinase; VEGF, vascular endothelial growth factor. (From Loeffek *et al.*, 2001.)

contact points, with tight tethering of resting keratinocytes to the dermis. This results in focal cleavage of the collagen matrix by MMP-1 bound to α2β1 integrin. The degraded collagen then denatures, which weakens the adhesion to the matrix and permits keratinocyte migration. Thus, α2β1 integrin, MMP-1, and collagen all coordinate together to drive and regulate migrating keratinocyte during reepithelialization.

The interactions of MMPs with proteoglycans provide another example of specialized function. Proteoglycans are large negatively charged molecules, which contain either heparan sulfate or chondroitin sulfate glycosaminoglycan chains covalently linked to a protein core. They are an important class of cell surface and ECM molecules that regulate the activation and activity of MMPs. Several MMPs bind to tissues by associating with glycosaminoglycans, with MMP-7 binding the most tightly.[8] The hyaluronan receptor, CD44, is covered with sulfated glycosaminoglycans, and MMP-7 binds to heparan sulfate proteoglycans that is close to epithelial cells and the underlying basement membrane; this binding enhances is enzymatic activity.[8] Because MMP-7 is an important regulator of many proteolytic events at the cell surface, the interactions between CD44 and MMP-7 may provide a mechanism for localizing the enzyme. MMP-2, -9, and -13 also bind to heparan sulfate in the extracellular space, a process that sequesters these enzymes in latent form. This may facilitate the ability of cells to regulate the activation of these enzymes, depending on the state of the microenvironment, that is, whether it is inflamed or not.

Endocytosis of pericellular MMPs is a way to suppress pericellular proteolysis, and low-density lipoprotein receptor-related protein (LRP) is an important mediator of this function. LRP is a heterodimeric endocytic receptor for many proteins, including α2-macroglobulin-proteinase complexes, and proteinase–inhibitor complexes, which block MMP activity.[8] ProMMP-2 has a low affinity to LRP, but when complexed with thrombospondin or TIMP-2, is efficiently endocytosed by LRP. ProMMP-9 is internalized as a proMMP-9–TIMP-1 complex, and MMP-13 is endocytosed when it binds to an as-yet unidentified 170-kDa receptor and then the enzyme is targeted for intracellular degradation.[8]

In summary, several studies have shown that secreted MMPs are recruited to the cell surface, where they interact with cell surface receptors or pericellular macromolecules. These interactions set in motion a complicated cascade, which carefully controls proteolysis, often in response to a biochemical or hormonal gradients present in the cellular microenvironment. While these interactions may increase expression and activity of MMPs, they also have the equally important function of downregulating MMPs under the appropriate circumstances. When these intricate mechanisms become dysregulated, the activities of MMPs, secrete or membrane-bound, can become pathologic.

4.4 MMPs in Exosomes

Exosomes are microvesicles of 30–120 nm, which are derived from endocytic membranes, that is, vesicles that form as a result of invagination of the plasma membranes (Fig. 4.5).[11] They are released from the cell by membrane fusion between the vesicular bodies and

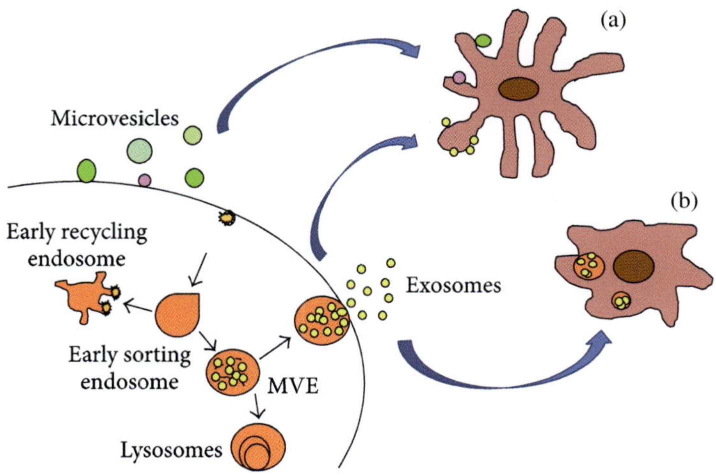

Figure 4.5. Exosomes are released from host cell and taken up by recipient cells. Exosomes are generated in host cell by merging microvesicular exosomes (MVEs) with the cell membrane and releasing into the extracellular space. These exosomes can be fused with the plasma membrane (a) or be internalized (b) by recipient cells. (From Lin *et al.*, 2015.)

plasma membrane. These vesicles contain a wide range of proteins, nucleic acids, and lipids and are found in many, if not all, bodily fluids. Initially, it was thought that exosomes were "garbage bags" for cells to get rid of unwanted constituents.[11] More recently, however, it has become clear that they carry the various molecular constituents of the cells from which they were derived: proteins, lipids, mRNAs, and microRNAs (miRNAs), and that they have important functions as modulators of cell behavior.[9,11–13] Because they originate from endosomes, exosomes contain specific marker proteins that are involved in the endosomal sorting complexes required for transport (ESCRT) (e.g., TSG101, Alix) and in transport and fusion (e.g., Rab11, Rab7, Rab2, and annexins). Other markers expressed in or on exosomes include tetraspanins (CD81, CD63, CD9), heat shock proteins (HSC70 and HSP90), and cytoskeletal proteins (e.g., actin and tubulin). Further, characterizing the molecular components in exosomes from normal cells has revealed the presence of various cell-type proteins. Some examples: exosomes isolated from antigen presenting cells harbor major histocompatibility complex II (MHCII) on their surface, while those from T-cells carry the TCR/CD3/zeta complex.[13] These cell-specific proteins target exosomes to specific recipient cells either by interaction with cell-surface adhesion molecules or through interaction with cell-surface heparan sulfate proteoglycans.[13] Alternatively, exosomes can be taken up by recipient cells through lipid-dependent endocytosis, where membrane lipids of the exosomes fuse with the plasma membrane of target cells.[13] Internalization of exosomes by recipient cells can, therefore, be a cell type-dependent event, providing a mechanism for imparting crucial instructions for the recipient cell.

Indeed, exosomes have important roles in cell-to-cell communication and they influence both normal physiological and pathological processes. In general, the biological functions of exosomes are still incompletely understood.[9,11,12] However, recent proteomics analyses of exosomes from cell lines and body fluids provide some clues about their biological significance and importance in human diseases.[11,12] MMPs are found in exosomes, where their transfer to recipient cells may alter the cell surface and mediate ectodomain shedding.[12]

One obvious example of MMPs in exosome-like vesicles extruded from the cell is the granules in polymorphonuclear leukocytes and mast cells. These cells store MMPs (and other proteinases) in vesicles, which are released into the extracellular microenvironment when the cells are activated, often by inflammatory stimuli.[9] Since both exosomes and MMPs are ubiquitous, it is not surprising that MMPs have been detected in vesicles found in endothelial cells (ECs), chondrocytes, and various cancer cells,[9] and that the MMPs in exosomes shed by cells are frequently linked to matrix remodeling and cellular invasion by normal cells. Activation of ECs with angiogenic factors such as fibroblast growth factor-2 (FGF-2) and/or vascular endothelial growth factor (VEGF) causes rapid release of exosomes containing MMP-2, MMP-9, MT1-MMP, and tissue inhibitors of metalloproteinases (TIMPs). The rapidity of the response suggests that the vesicles are stored in intracellular storage compartments. MMPs in exosomes are in both pro- and active forms, and when they are taken up by ECs, the formation of capillary-like structures *in vitro* increases, as does EC invasion.[9] Similarly, cultures of chondrocytes derived from growth plate cartilage produce matrix vesicles that contain pro- and active MMP-2 and MMP-3, and TIMPs, and treating the cells with the vitamin D metabolite 1,25-$(OH)_2D_3$ increases MMP activity. Vesicles from chondrocytes isolated from growth zones have higher levels of MMPs than chondrocytes from resting zones, suggesting that MMPs are contributing to remodeling of the ECM at the hypertrophic cell zone in the growth plates of long bones.[9]

However, by far and away, the most interest in exosomes has been in cancer biology.[9,12,13] Tumor-specific exosomal proteins have been implicated in tumor progression and have been referred to as "a message delivery system for tumor progression".[13] In melanoma cells, MMP-2 and MMP-9 have been found in vesicular organelles located along the microtubular network.[9] These vesicles also contained the motor proteins, kinesin, and α-tubulin. Treating the cells with a microtubule-interfering drug reduced the secretion of MMP-2 and MMP-9, supporting the concept that some cell types can store these MMPs intracellularly in exosomes. Since kinesin may actively propel the exosomes along microtubules towards the plasma membrane, shedding of these vesicles with MMPs as part of their cargo may provide one mechanism for directional proteolysis during cell migration and invasion.[9]

Exosomes found in ascites fluid from ovarian carcinoma contain MMPs, along with other classes of proteinases: pro- and active forms of MMP-2, MMP-9, and MT1-MMP, active urokinase-like plasminogen activator, and urokinase-like plasminogen activator receptor. Patients with later stage cancers had more vesicles, with more active enzymes, than did patients with earlier stage disease, emphasizing the important concept of cooperativity among classes of proteinases in mediating tumor progression. Further, these vesicles stimulated the invasion of ovarian cancer cell cultures through Matrigel®, a process that was blocked if inhibitors of MMPs or serine proteases were added.[9]

Finally, gastrointestinal stromal tumor (GIST) cells secrete exosomes, which contain oncogenic KIT. Their uptake by normal surrounding stromal smooth muscle cells leads to an increase in downstream signaling through the AKT and mitogen activated protein kinase (MAPK) signaling and phenotypic changes in cells. These changes include changes in morphology, expression of tumor-associated markers such as vimentin and smooth muscle actin, enhanced expression of MMP-1, and increased tumor cell invasion (Fig. 4.6).[13] This study emphasizes host-tumor cells interactions in tumor progression and also documents the critical role of exosome-mediated signaling in the tumor microenvironment, where it acts as a positive feedback loop that facilitates tumor invasion.

In summary, exosomes are increasingly recognized as essential components of a cell–cell communication network. Given that there appears to be considerable specificity in the targeted uptake of exosomes by recipient cells, the possibility arises that pharmacological manipulation of this message delivery system may represent a promising therapeutic approach.[13]

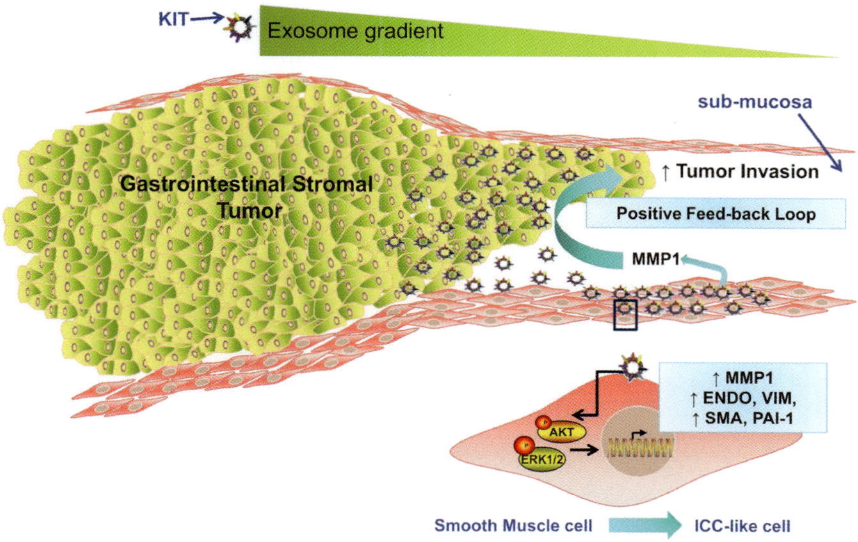

Figure 4.6. Proposed model of tumor-stromal positive feedback loop mediated by oncogenic KIT-bearing tumor exosomes in the regulation of tumor invasion. GIST cells secrete a gradient of exosomes carrying mutant KIT, which after internalization by surrounding smooth muscle cells activate downstream signaling pathways of KIT (e.g., AKT and MAPK pathways) and induce an enhanced expression of endoglin, vimentin (VIM), smooth muscle actin (SMA), and plasminogen activator inhibitor-1 (PAI-1) in the recipient cells, resembling an ICC-like phenotype. This tumor-stromal interaction creates a positive feedback loop in which tumor-derived exosome-mediated signaling in stromal cells increases MMP1 secretion. In turn, tumor cells utilized MMP1 to invade the submucosa. This model describes a previously unreported mechanism by which tumor-derived exosomes can modulate their host microenvironment and promote local invasion and potentially distant metastasis. (From Atay and Godwin, 2014.)

4.5 Intracellular Localization of MMPs

A specific proteolytic enzyme usually has a dedicated function in an intracellular, extracellular, or intranuclear location, but not in all three. However, as more and more studies with MMPs have accrued, both secreted and membrane-bound MMPs have been localized to cytosolic and to nuclear sites.[4,9,14] Although a report of MMPs localized in the cytoplasm or nucleus might be dismissed as "sloppy science" and/or errant protein targeting, careful biochemical analyses and the identification of a nuclear localization signal(s) (NLS) of some MMPs suggest that these alternate localizations are an important aspect of MMP behavior (Fig. 4.7).[4,9,14]

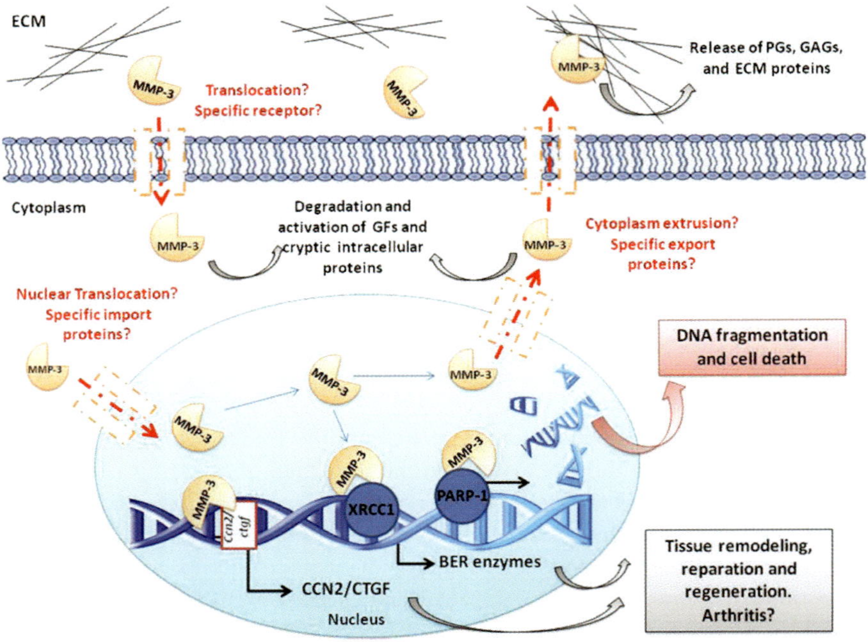

Figure 4.7. Schematic representation of the multiple roles of extracellular, intracellular, and intranuclear MMP-3 in chondrocytes. Based on numerous previous reports, stromelysin-1 (MMP-3) has been recognized as an MMP functioning as a regulatory proteinase degrading substrates present in extra- and intracellular milieu, but also modulating gene expression and cleaving peculiar proteins in the nuclei. In fact, the canonical presence of MMP-3 in extracellular space (by the well-known secretion pathway mainly related to the N-terminal signal peptide of MMPs) is associated to the enzymatic degradation of types IV and IX collagens, laminin, fibronectin, elastin, and proteoglycans. After the translocation from ECM to cytosol (through a putative receptor), the degradation of nonmatrix cytosolic proteins inside the cells have been also characterized, mainly linked to the activation of several biological precursors like proTNF-α, proIL-1β, IGFBPs, myelin basic protein, proMMP-1, and so on. After translation or via the protein-mediated nuclear transportation, the novel roles of MMP-3 have been identified as gene regulator (after the binding of MMP-3 with the TRENDIC motif for transregulation of the *CCN2/CTGF* gene) and as nuclear proteinases cleaving peculiar proteins (such as PARP and XRCC1). These findings support a role for MMP-3 in apoptosis, gene regulation, and tissue remodeling. (From Mannello and Medda, 2012.)

The "cytoplasm" is defined as the portion of a cell that is enclosed by the plasma membrane, and thus it contains the cytosol and the organelles within it.[14] When determining the intracellular localization(s) of MMPs, simple immune-staining is not definitive since it may not discriminate between cytosol and organelles. Further, proMMPs are normally present intracellularly when as they are synthesized in the endoplasmic reticulum (ER) and processed for secretion in the Golgi. Therefore, accurate cytoplasmic documentation of an activated MMP within the cytoplasm requires careful subcellular fractionation and detailed biochemical evidence.[14] However, even with careful biochemistry and evidence of proteolytic cleavage of substrate *in vitro*, a cautionary note remains as to whether the novel substrate tested *in vitro* is, in reality, a true biological substrate with a "real" role in physiology and/or pathology.[2,14]

Nonetheless, several reports have consistently documented the cytoplasmic location and function of MMPs, where they have a growing range of substrates, with potentially important functions. Intracellular MMP-1 has been demonstrated in several types of cells, including glia cells, epithelial cells, and fibroblasts.[9] Early in apoptosis, both the proMMP-1 and active MMP-1 were colocalized with mitochondria that were clustered near the nucleus. During the later stages of apoptosis, MMP-1 collected around the nucleus and nuclear fragments, suggesting a possible role for this enzyme in breaking down the nuclear envelope. In addition, the levels of intracellular MMP-1 varied during the cell cycle, with the highest levels seen during M phase, implicating MMP-1 in cell growth.[9]

Active MMP-2 has been demonstrated intracellularly in myocytes, where it is most likely activated via peroxynitrite inside cardiac myocytes undergoing injury from ischemia-reperfusion injuries.[9,14] Ischemia/reperfusion is part of acute myocardial injuries, which are characterized by a loss of muscle contractility during the postischemic reperfusion phase. The presence of active MMP-2 in injured myocytes suggests that the loss of contractility results from the ability of MMP-2 to cleave the contractile proteins, TnI and the cytoskeletal protein, α-actinin.[9] In yet another example, active MMP-3 is found intracellularly, where it helps to mediate apoptosis in dopaminergic neurons. During apoptosis, the proform of MMP-3 was proteolytically cleaved active MMP-3 by a serine proteinase.[9] The finding that in the absence of intracellular MMP-3 activity, the cells were protected from apoptosis supports a role for MMP-3 in this process.

4.6 Nuclear Localization

Many MMPs, including MMP-2,-3,-9,-13,-26 and MT1-MMP, have been found in the nucleus of several types of cells (including heart myocytes, brain neurons, ECs, fibroblasts, and hepatocytes) where they may cleave nuclear matrix proteins.[4,9,14] The nuclear matrix is surrounded by a nuclear lamina and a double-membrane nuclear envelope, and therefore, requires active mechanisms for nuclear import and export in order to shuttle large macromolecules in and out of the nucleus.[14] Consequently, if MMPs enter or exit the nucleus, their amino acid sequence must harbor specific signals, termed NLSs, or nuclear export signals (NESs) (Fig. 4.8).

Although the mechanisms of nuclear translocation of MMPs are not well understood, several mechanisms have been proposed. MMP-2 has a classical nuclear localization sequence close to the C-terminus, which could result in trafficking to the nucleus.[4,9,14] Similarly, a nuclear signaling sequence is found in the catalytic domain of MMP-3, and it is required for this enzyme to be translocated to the nucleus. Interestingly, full-length MMP-3 is not found in the nucleus, indicating that some molecular processing is needed to expose the NLS for nuclear transport. For MT1-MMP, it has been suggested that a caveolae-mediated mechanism of endocytosis mediates nuclear translocation to perinuclear regions.

As reports of intranuclear MMPs continue to accumulate, some themes about their possible functions are beginning to emerge.[4,9,14] Nuclear localization of MMP-2, MMP-9, and MMP-13 has been associated with DNA damage and apoptosis following ischemic injury in pulmonary ECs, in rat brain neurons, and in cardiac myocytes. In all three of these cell types, nuclear enzymatic activity correlated with the ability of MMPs to proteolytically interfere with proteins involved in DNA repair. In the ECs, MMP-2 activation can

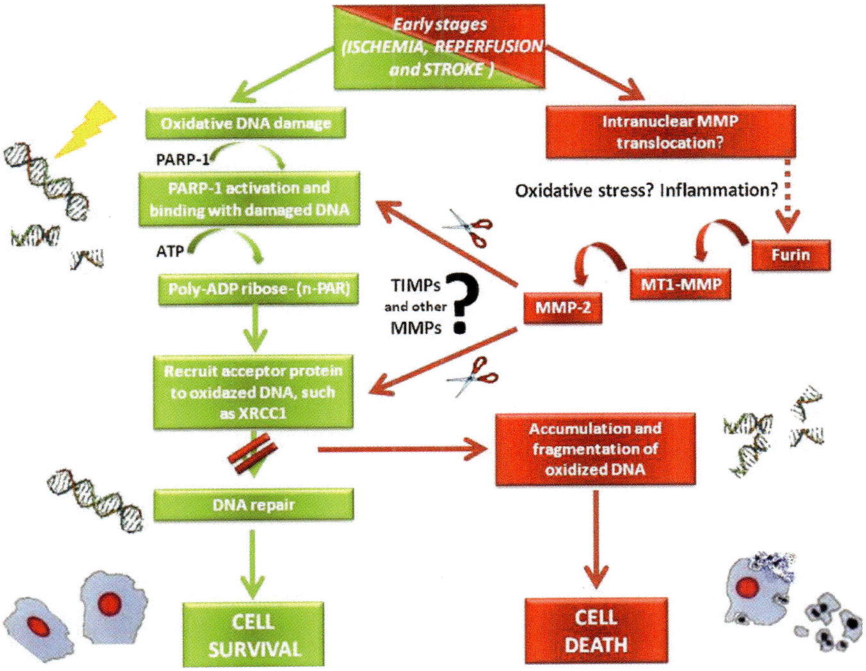

Figure 4.8. Hypothetical diagram of the events linking intranuclear MMPs and oxidative DNA damage in brain cells during ischemia, reperfusion, and stroke. In ischemic brain tissue (shortly after the injury), some nuclear MMPs are able to cleave nuclear proteins, which facilitate accumulation of oxidative damaged DNA at an early stage after an ischemic insult. In particular, MT1-MMP colocalized with MMP-2 in the ischemic nuclei and in nuclear extracts, jointly with the expression of furin, an activator of MT1-MMP. Ischemic nuclear extracts contained a significant increase of MMP-dependent cleavage products of both poly-ADP-ribose polymerase-1 (PARP-1) and X-ray cross-complementary factor 1 (XRCC1), nuclear proteins that play crucial roles in DNA fragmentation, DNA base excision repair, and cell apoptosis. The degradation of these nuclear proteins via MMP-2 and MMP-9 activated by furin-enhanced MT1-MMP activity during ischemic insult enhanced nuclear accumulation of oxidative DNA damage and leaving the reperfused ischemic rat brain prone to apoptosis. (From Mannello and Medda, 2012.)

be induced by reactive oxygen and nitrogen species produced by cigarette smoke. In rat brain neurons, MMP-2 and MMP-9 activity is associated with DNA fragmentation, which was reduced by MMP inhibitors or by MMP-specific antibodies. Nuclear MMP-13 was increased in neural cells after cerebral ischemia in both rats and humans, and increased translocation to the nucleus occurred with deprivation of oxygen and glucose. MT1-MMP and its activator, furin, were also detected in the nucleus of the ischemic rat brain neurons, suggesting that intracellular activation of MMP-2 may occur through MT1-MMP.

Active MMP-3 in the nucleus has a novel role as a transcription factor, and therefore, has received considerable attention.[4,9,14] In cultured chondrocytes *in vitro,* and in the normal and osteoarthritic chondrocytes *in vivo,* MMP-3 binds to a transcription enhancer sequence in the promoter of connective tissue growth factor (*CCN2/CTGF*) gene and activates its transcription. CTGF promotes chondrocyte growth and matrix deposition. Notably, both pro- and active MMP-3 could activate the CCN2/CTGF promoter, suggesting that this novel function may be independent of enzymatic activity. Further, both the hemopexin and the catalytic-hinge regions of MMP-3 activated the promoter, while the prodomain and the hinge-region alone could not. A DNA-binding domain was found in the hemopexin domain, as an antibody directed against the MMP-3 hemopexin domain antibody inhibited protein–DNA interactions. Interestingly, the hinge region contains proline-rich sequences found in some transcription factors, and thus may provide an explanation for this mode of transcriptional activation. However, catalytically dead MMP-3 mutants could also activate transcription, although at a lower level, suggesting that MMP-3 regulates the CCN2/CTGF promoter activity by two different mechanisms. One may involve proteolytic processing of one or several nuclear proteins, while the other is independent of the proteinase activity and involves the hemopexin domain. Future studies may determine if one or the other of these putative mechanisms is dominant and/or whether they act in concert.

To summarize, the discovery of MMPs in localized compartments within the cell and in the pericellular environment represents an important, but underappreciated, aspect of MMP functions.

These discoveries open a new field in the cell biology and biochemistry of MMPs. The novel compartmentalization of cytoplasmic and nuclear MMPs illustrates their important roles in diverse physiological and pathological cellular processes, where MMPs are either expressed constitutively or induced, and where they have regulatory functions. These relatively newly described roles for MMPs emphasize the importance of the mechanisms controlling their expression and enzymatic activities. Further investigations into intracellular/intranuclear sorting and trafficking are important areas for future studies, as is the careful identification of specific nuclear and cytoplasmic substrates. Results of these kinds of studies may provide a foundation for designing specific agents that can block MMPs both under physiological and pathological conditions.

4.7 MMPs as Signaling Molecules

MMPs can be defined as "signaling molecules" when they are involved in mediating the interactions of a ligand with its receptor, with the subsequent initiation of signal transduction pathways, and downstream modification of gene expression and cell behavior. MMPs can activate receptors in several ways. First, when MMPs liberate growth factors, cytokines, and hormones embedded in the ECM, these ligands are now free to bind to their cognate receptors with subsequent initiation of signal transduction cascades. In this instance, MMPs are indirect and almost serendipitous participants in modulating cell behavior: they happen to be at the right place and at the right time. Second, abundantly high levels of activated MMPs can proteolytically cleave either receptors on cell membranes or their ligands, and consequently interfere with the ability of the cells to respond to a signal since the receptor is now essentially destroyed or rendered inactive. Third, proteolytic cleavage of a class of G protein-coupled receptors (GPCRs) by MMPs results in particular targeted activation of signaling cascades with profound effects on cell behavior. Specifically, these GPRCs are protease activated receptors (PARs), of which there are four, with PAR-1 being the most ubiquitous; both

MMP-1 and MMP-13 can cleave PAR-1, resulting in activation of signaling pathways (Fig. 4.9).[15]

1. *Liberation of growth factors and cytokines.* While the liberation of growth factors and cytokines embedded in the ECM is well known, the second and third roles a signaling molecules are more subtle. MMPs are the "perpetrators" of signaling when their proteolytic "skills" result in enzymatic modification of signaling molecules, sometimes with less than beneficial results. For example, MMPs can liberate the matrix-bound ligand, VEGF A, so that it can bind to its receptor. However, MMP-9 (along with

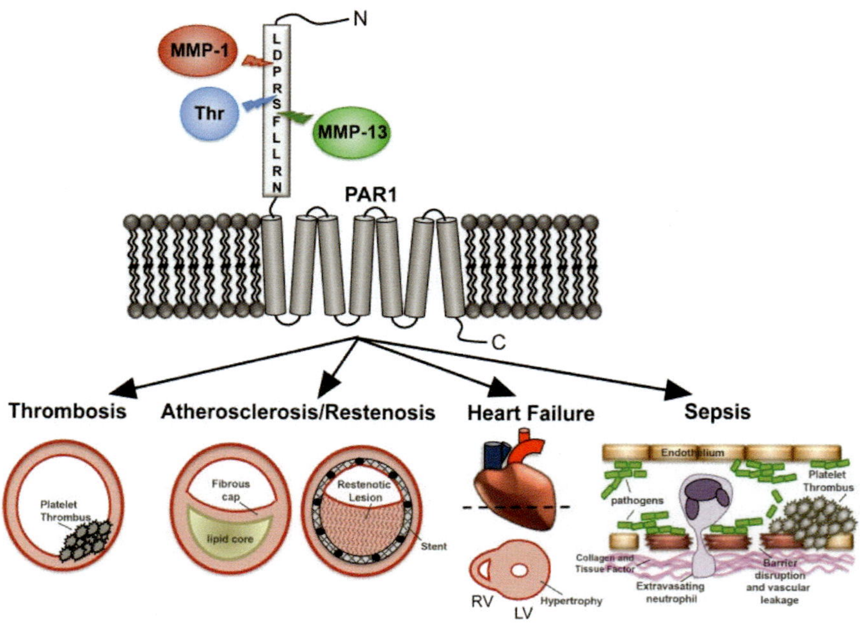

Figure 4.9. MMP-PAR1 signaling in vascular diseases. PAR1 senses a diverse milieu of extracellular proteases and subsequently relays that information to influence cellular behavior and potentially exacerbate disease pathologies. The N-terminal extracellular domain (exodomain) of PAR1 is cleaved at a canonical site by thrombin and noncanonical sites by MMP-1 and MMP-13. Various signaling outputs can lead to platelet thrombosis, atherosclerosis, in-stent restenosis, heart failure, and sepsis. (From Austin *et al.*, 2013.)

plasmin) can also cleave the C-terminal domain of VEGF, which hastens its degradation and hinders the recruitment of a new vascular supply (Fig. 4.10).[16] Given the potential for abundant amounts of MMPs and of VEGF A in normal physiology and in pathological conditions, the ability of proteolytic "sculpting" of VEGF A by MMPs would seem to be a potentially potent mechanism for influencing cell behavior and possibly, disease outcome.

2. *Proteolysis of receptors.* Modulation by MMPs of ligand/receptor interactions through proteolysis of receptors can have several downstream consequences for signal transduction and

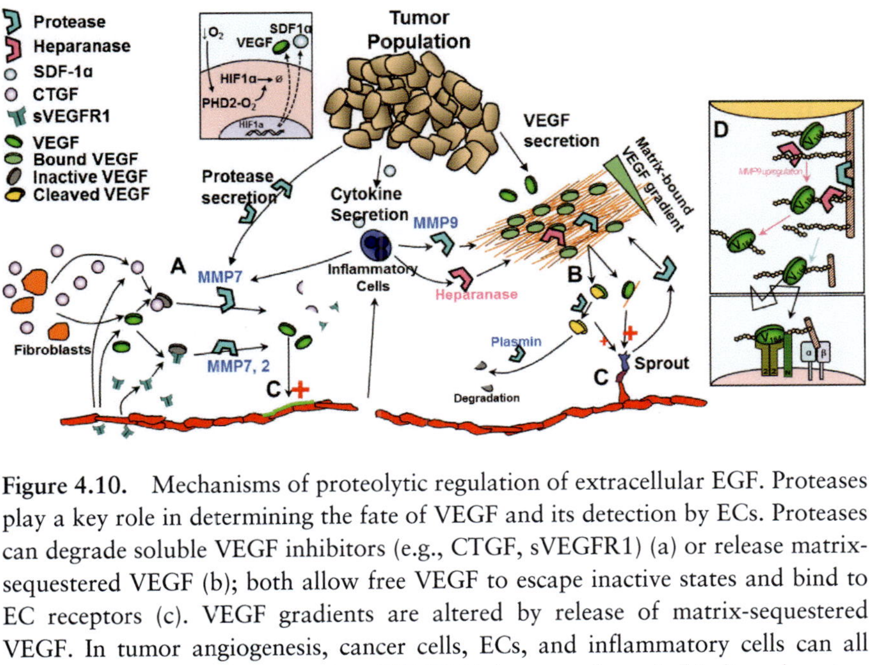

Figure 4.10. Mechanisms of proteolytic regulation of extracellular EGF. Proteases play a key role in determining the fate of VEGF and its detection by ECs. Proteases can degrade soluble VEGF inhibitors (e.g., CTGF, sVEGFR1) (a) or release matrix-sequestered VEGF (b); both allow free VEGF to escape inactive states and bind to EC receptors (c). VEGF gradients are altered by release of matrix-sequestered VEGF. In tumor angiogenesis, cancer cells, ECs, and inflammatory cells can all contribute to proteolytic activity. VEGF164 has two heparin-binding domains; cleavage of either domain results in a VEGF164/113 intermediate (b), with lower overall affinity for the ECM. Subsequent cleavage of the second domain results in freely diffusing VEGF113. Some MMPs can cleave VEGF bound to heparin/HSPGs (e.g. MMP-3) while others cannot (e.g. MMP-9). Heparanase activity on cell-surface HSPGs can lead to upregulation of MMP-9 leading to cleavage of both GAG chains and core protein (d) and enhanced signaling at the cell surface MMPs and VEGF. (From Vempati *et al.*, 2014.)

cell behavior. One important example of receptor proteolysis has formed the basis for a hypothesis that explains the comorbidities associated with "metabolic syndrome". This is a cluster of clinical entities of hypertension, diabetes, reduced capillary density (rarefaction), profound obesity, and immune suppression.[17–19] High levels of MMPs have been implicated in this syndrome with changes in the arterial vasculature. The changes begin with endothelial dysfunction and lead to micro- and macrovascular complications, and eventually, remodeling of the basement membrane (Fig. 4.11).[17] These changes also promote thrombosis, followed by leukocyte activation, oxidative stress and, eventually, to increases in MMPs expression. While elevated MMPs expression in metabolic syndrome is well-documented, two lines of evidence support a role of MMPs in mediating hypertension: (1) MMPs are expressed in conditions where hypertension is a risk factor, such as stroke, atherosclerosis, renal disease, heart disease, and arterial aneurysm, and (2) MMPs are involved in cardiac hypertrophy and its transition into heart failure in hypertension.

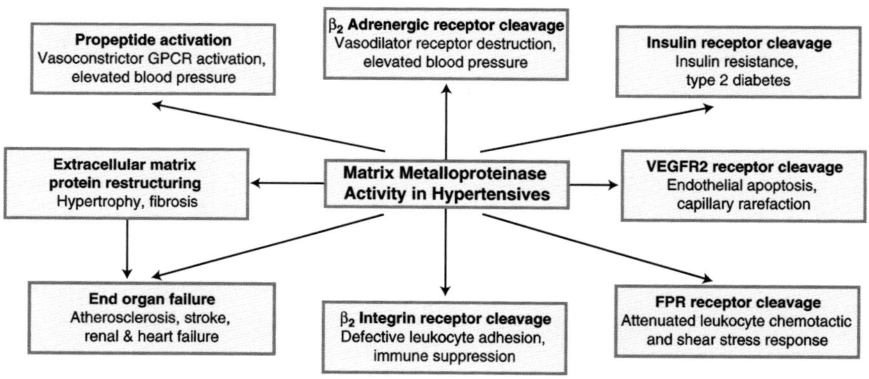

Figure 4.11. MMPs and metabolic syndrome. Schematic summary of evidence supporting a hypothesis for involvement of MMP in hypertension, ventricular wall hypertrophy, insulin resistance, capillary rarefaction, attenuated leukocyte adhesion and shear stress response, and eventual organ failure. FPR, formyl peptide receptor; GPCR, G protein–coupled receptor; VEGFR2, vascular endothelial growth factor receptor 2. (From Schmid-Schönbein, 2012.)

In addition, unbridled MMP activity cleaves other important receptors in this syndrome, including the β_2 adrenergic receptor. Binding of agonist usually stimulates vasodilation but in the arterioles of the experimental system of the spontaneously hypertensive rat, the ability of the β_2 adrenergic receptor to increase vasodilation is compromised, because its extracellular domain has been cleaved off by MMPs.[17] The insulin receptor-α also shows evidence of cleavage of its extracellular domain by MMPs, and this cleavage is associated with a reduced ability to transport glucose across its membranes. The fact that chronic inhibition of MMP enzyme activity reduces elevated blood glucose and glycosylation supports a role of MMPs in mediating at least some of the pathogenesis of this syndrome. Similarly, cleavage of the extracellular domain of the β_2 adrenergic receptor is seen in other models of insulin resistance, providing additional support for the ability of MMPs to contribute to the aberrant regulation of glucose metabolism.

Still another membrane receptor in the hypertensive rat is also cleaved.[17] The extracellular domain of the transmembrane G-protein coupled receptor (GPCR), formyl peptide receptor, is normally found on circulating leukocytes, but is less abundant and has a reduced response to chemotactic peptides in the diseased rat. In addition, this receptor serves as a mechano-sensor in retracting pseudopod projections in circulating leukocytes in response to fluid shear stress. Compared to normal rats, its extracellular domain is cleaved in the hypertensive rat, and there is an increase in the number of circulating leukocytes with pseudopods, which increases both capillary and hemodynamic resistance. It seems, then, that multiple signaling components are affected by MMP proteolysis in metabolic syndrome. The novel hypothesis for the role of MMP-1 directly links the increase in the level of MMPs in this syndrome with some of its pathological features, thereby helping to provide a molecular mechanism to explain a wide range of clinical findings.

3. *MMPs and the GPCR, PAR-1.* GPCRs are seven-transmembrane receptors, whose activation results in a large diversity of cellular responses, which are mediated through various G-protein and

non–G-protein pathways (Fig. 4.9).[15,20,21] Although protease-activated receptors (PARs) are members of the GPCR superfamily, they have a unique mechanism of activation: proteolytic cleavage.[15,20] Although four PARs have been identified, PAR-1 is the most ubiquitous and the best studied.[15,20,22] PARs are expressed by several types of cells, including ECs, platelets, activated fibroblasts, and some types of tumor cells. PAR receptors are unique in that their ligand is embedded and masked N-terminally under resting conditions.[15,20] PAR-1 was initially identified as the first receptor for thrombin,[23] with the subsequent discovery of PAR2, PAR3, and PAR4.[23] The serine protease, thrombin, is the classic activator of PAR-1, and when the ligand is exposed by proteolytic cleavage, it binds to the extracellular active site and activates PAR-1 intramolecularly. Activated PAR initiates signaling through MAPK pathways with downstream consequences that alter cell morphology and behavior.

After the initial demonstration that thrombin could cleave PAR1 on human platelets, other members of the serine protease family were shown to cleave PARs, including plasmin, activated protein C, thrombocytin, platelet-activating enzyme from *Bothrops jaraca* (PA-BJ), factor Xa, factor VIIa, kallikreins, cathepsin G, trypsin, matriptase, and tryptase.[15] Since PARs are expressed on nearly all cell types in the blood vessel wall and blood, with the exception of red blood cells, and since PARs are activated by thrombin, these receptors have a vital role in the vasculature.[15] Specifically, PAR1 is expressed on the surface of ECs, smooth muscle cells, platelets, neutrophils, and macrophages.[15] Thrombin activation of PAR1 stimulates platelet aggregation, cell proliferation, changes in cell shape and adhesion, chemokine production, and cell migration.

In 2005, Kuliopulos and colleagues reported that MMP-1 could also cleave and activate PAR1, leading to unique patterns of signal transduction and gene expression that are distinct from that of thrombin.[15,24,25] This cleavage occurs at a noncanonical site,[15] which may explain the differential signaling pattern when compared to thrombin, perhaps resulting from distinct conformational changes that occur after proteolytic cleavage. Of importance, another collagenase, MMP-13, also cleaves PARs, at yet still a different cleavage site (Fig. 4.9).[15] MMP-13 cleavage of PAR occurs on cardiac fibroblasts and cardiomyocytes following β-adrenergic receptor stimulation. Consequently, it has been suggested that this signaling pathway results in pathologic activation of downstream signaling events that contribute to heart failure.[26]

The MMP-1/PAR1 signaling axis has a prominent role in the progression of several cancers, including ovarian, lung, and melanoma.[15,21,27,28] This role is distinct from that of the degradation of interstitial collagen, which, in cancer biology, frees the tumor cells to escape from the primary site. MMP-1/PAR-1 signaling acts on the tumor cells and on adjacent stromal cells to increase expression of a host of genes involved in cell behavior and pathological outcome.[21,25,29] Importantly, it is likely that thrombin and MMP-1 cooperate in the tissue microenvironments to additively or synergistically promote both tumor progression and inflammation.[21]

4.8 Conclusion

It is abundantly clear that MMPs are "not for matrix anymore". In fact, there may be more nonmatrix substrates proteolytically modified by MMPs than matrix proteins. Although MMPs are traditionally found in the extracellular milieu, their presence in the pericellular microenvironment of the secreted MMPs as well as membrane-bound is well-documented. They are associated with complex activities that direct cell movement and behavior, liberate growth factors, hormones and cytokines from matrix molecules, and activate or inhibit cellular receptors with their enzymatic activities. Intracellularly, they may have specialized functions in cytoplasmic organelles as well as in the nucleus, where several MMPs appear to be involved in DNA damage and apoptosis. Their presence in secreted vesicles (exosomes) allows them to be transported to new cells, where they are taken up and can discharge their functions either pericellularly or intracellularly. Finally, the ability to two collagenases, MMP-1 and MMP-13, to proteolytically cleave the GPCR, PAR1, assigns unique roles to these proteinases as powerful signaling molecules, which can activate MAPK pathways and modulate gene expression, with dramatic changes in cell behavior. No wonder, then, MMPs should now be viewed as major controllers of homoeostasis in the extracellular and systemic environments. By acting upstream in information pathways through the activation or inhibition of signal transduction pathways, MMPs can profoundly control cell function. While their original role of MMPs as "emperors" of connective tissue matrix destruction remains intact and undisputed, it is essential to recognize that this represents just one aspect of the intracellular/extracellular environments under the portfolio of MMP activities.

References

1. McCawley, LJ and LM Matrisian (2001). Matrix metalloproteinases: they're not just for matrix anymore! *Current Opinion Cell Biology*, 13, 534–540.
2. Butler, GS and CM Overall (2009). Updated biological roles for matrix metalloproteinases and new 'intracellular' substrates revealed by degradomics. *Biochemistry*, 48, 10830–10845.
3. Rodríguez, D, CJ Morrison and CM Overall (2010). Matrix metalloproteinases: what do they not do? New substrates and biological roles identified by murine models and proteomics. *Biochimica et Biophysica. Acta — Molecular Cell Research*, 1803, 39–54.
4. Mannello, F and V Medda (2012). Nuclear localization of matrix metalloproteinases. *Progress in Histochemistry and Cytochemistry*, 47, 27–58.
5. Koziol, A, M Martín-Alonso, C Clemente, P Gonzalo and AG Arroyo (2012). Site-specific cellular functions of MT1-MMP. *European Journal Cell Biology*, 91, 889–895.
6. Lin, HC *et al.* (2001). Matrilysin cleavage of corneal collagen type XVIII NC1 domain and generation of a 28-kDa fragment. *Investigative Ophthalmology Visual Science*, 42, 2517–2524.
7. Ma, DHK *et al.* (2007). Generation of endostatin by matrix metalloproteinase and cathepsin from human limbocorneal epithelial cells cultivated on amniotic membrane. *Investigative Ophthalmology and Visual Science*, 48, 644–651.

8. Murphy, G and H Nagase (2011). Localizing matrix metalloproteinase activities in the pericellular environment. *FEBS Journal*, 278, 2–15.
9. Hadler-Olsen, E, B Fadnes, I Sylte, L Uhlin-Hansen and JO Winberg (2011). Regulation of matrix metalloproteinase activity in health and disease. *FEBS Journal*, 278, 28–45.
10. Löffek, S, O Schilling and CW Franzke (2011). Series 'matrix metalloproteinases in lung health and disease': biological role of matrix metalloproteinases: a critical balance. *European Respiratory Journal Off. Journal of European Socirty Clinical Respiratory Physiology*, 38, 191–208.
11. Lin, J *et al.* (2015). Exosomes: novel biomarkers for clinical diagnosis. *The Scientific World Journal*, 2015.
12. Shimoda, M and R Khokha (2013). Proteolytic factors in exosomes. *Proteomics*, 13, 1624–1636.
13. Atay, S and AK Godwin (2014). Tumor-derived exosomes: a message delivery system for tumor progression. *Communicative and Integrative Biology*, 7, 1–6.
14. Cauwe, B and G Opdenakker (2010). Intracellular substrate cleavage: a novel dimension in the biochemistry, biology and pathology of matrix metalloproteinases. *Critical Review in Biochemistry and Molecular Biology*, 45, 351–423.
15. Austin, KM, L Covic and A Kuliopulos (2013). Matrix metalloproteases and PAR1 activation. *Blood*, 121, 431–439.
16. Vempati, P, AS Popel and F Mac Gabhann (2014). Extracellular regulation of VEGF: isoforms, proteolysis, and vascular patterning. *Cytokine Growth Factor Reviews*, 25, 1–19.
17. Schmid-Schönbein, GW (2012). An emerging role of degrading proteinases in hypertension and the metabolic syndrome: autodigestion and receptor cleavage. *Current Hypertension Reports*, 14, 88–96.
18. Berg, G, V Miksztowicz and L Schreier (2011). Metalloproteinases in metabolic syndrome. *Clinica Chimica Acta*, 412, 1731–1739.
19. Hopps, E and G Caimi (2012). Matrix metalloproteinases in metabolic syndrome. *European Journal of International Medicine*, 23, 99–104.
20. Traynelis, SF and J Trejo (2007). Protease-activated receptor signaling: new roles and regulatory mechanisms. *Current Opinion in Hematology*, 14, 230–235.
21. Eck, SM, JS Blackburn, AC Schmucker, PS Burrage and CE Brinckerhoff (2009). Matrix metalloproteinase and G protein coupled receptors:

co-conspirators in the pathogenesis of autoimmune disease and cancer. *Journal of Autoimmunity*, 33, 214–221.
22. Zigler, M, T Kamiya, EC Brantley, GJ Villares and M Bar-Eli (2011). PAR-1 and thrombin: the ties that bind the microenvironment to melanoma metastasis. *Cancer Research*, 71, 6561–6566.
23. Coughlin, SR (2005). Protease-activated receptors in hemostasis, thrombosis and vascular biology. *Journal of Thrombosis Haemostasis*, 3, 1800–1814.
24. Boire, A *et al.* (2005). PAR1 is a matrix metalloprotease-1 receptor that promotes invasion and tumorigenesis of breast cancer cells. *Cell*, 120, 303–313.
25. Blackburn, JS and CE Brinckerhoff (2008). Matrix metalloproteinase-1 and thrombin differentially activate gene expression in endothelial cells via PAR-1 and promote angiogenesis. *American Journal of Pathology*, 173, 1736–1746.
26. Jaffré, F, AE Friedman, Z Hu, N MacKman and BC Blaxall (2012). β-Adrenergic receptor stimulation transactivates protease-activated receptor 1 via matrix metalloproteinase 13 in cardiac cells. *Circulation*, 125, 2993–3003.
27. Lee, HJ, B Wall and S Chen (2008). G-protein-coupled receptors and melanoma. *Pigment Cell and Melanoma Research*, 21, 415–428.
28. Villares, GJ, M Zigler and MB Eli (2011). The emerging role of the thrombin receptor (PAR-1) in melanoma metastasis — a possible therapeutic target. *Oncotarget*, 2, 8–17.
29. Blackburn, JS, I Liu, CI Coon and CE Brinckerhoff (2009). A matrix metalloproteinase-1/protease activated receptor-1 signaling axis promotes melanoma invasion and metastasis. *Oncogene*, 28, 4237–4248.

5. Matrix Metalloproteinases (MMPs) in Homeostasis and Health: Can MMPs be Good for You?

Constance E. Brinckerhoff

5.1 Introduction

The answer to this question is a resounding "YES"! Matrix metalloproteinases (MMPs) have critical roles in homeostasis as they maintain normal remodeling of our cells and tissues. In the beginning, knowledge of the basal levels of MMP expression and the consequences of this expression were often deduced from experiments where "normal" cells and tissues were compared to tissues from patients with disease. Results from these studies revealed that under homeostasis, MMPs levels are generally low (see Table 3.1).[1–3] In this mode of "normal", their actions are often subtle and unnoticed or unrecognized. It was not until clinical trials with broad-spectrum MMP inhibitors (MMPIs) revealed unexpected and painful side effects that we began to see some of the less dramatic, but extremely important, actions of these enzymes.

Interestingly, punctuated amidst these normally low levels of expression are carefully controlled explosions of heightened MMP expression in situations such as wound healing and uterine resorption after childbirth, to give just two examples, of which there are many. What is so remarkable about these intense bursts of "normal" MMP activity is the exquisite orchestration of events: there is just enough MMP expression to get the job done the way it is supposed to be done, with the desired results of, for example, a healed wound (with nonscarring) and a uterus that has returned to its prepregnant state.

Documentation of these subtle remodeling and homeostatic activities of MMPs has led to the conclusion that our connective tissues (skin, bone, tendons, cartilage, etc.) are not static. Rather, they are dynamic organs and tissues that are constantly undergoing enzymatic modeling and remodeling, often in response to the activation of signal transduction and gene regulation pathways. Remarkably, these normal control processes know when to begin and when to end. In contrast, these mechanisms go awry in disease pathology, when MMPs expression becomes dysregulated and essentially out of control.[4] Given the ever-increasing portfolio of MMP activities, inside and outside of the cells, as enzymes and as signaling molecules, it becomes apparent that reigning in the behavior of these enzymes to appropriately control them for therapeutic regimes is a complicated and difficult process.

5.2 Experimental Systems and Normal Development

Perhaps one of the earliest demonstrations of the constitutive low levels of MMP expression in normal tissues came from experiments comparing MMP expression in diseases tissues to "normal" controls. In particular, synovium taken from connective tissues was studied. Tissue taken from joints of patients with rheumatoid arthritis was the subject of early investigations, since synovium from these patients had been prime source of material from which to purify human collagenase. Shortly thereafter, the "inducing principle", that is, the factor that was responsible for increasing collagenase in synovium from rheumatoid patients was identified as Interleukin-1β (IL-1β).[1,5] Other examples of these low MMP levels in normal tissues are normal fibroblasts versus those that have been stimulated with growth factors and cytokines, which express substantially higher levels (Fig. 5.1).[1,6,7] These findings suggest that, usually, in a healthy resting state, levels of MMP expression are low.

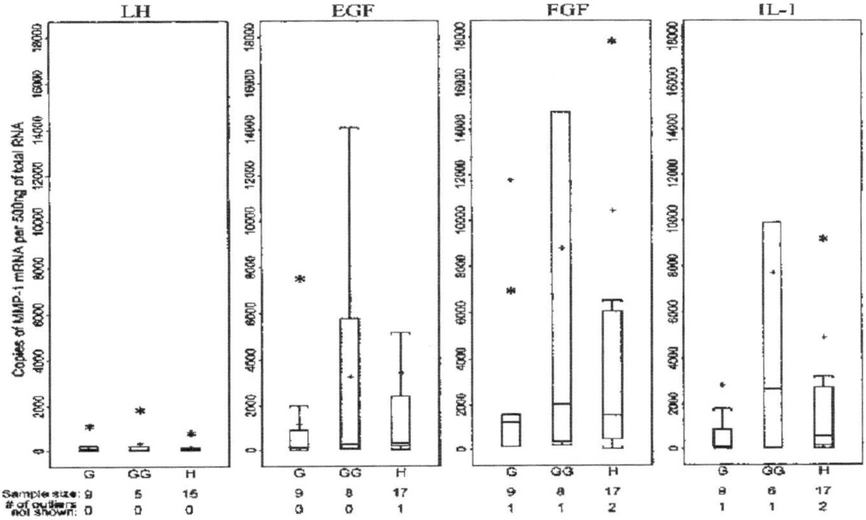

Figure 5.1. MMP-1 expression in response to growth factors and cytokine. Box plots of the distribution of MMP-1 expression levels in *1G* homozygous (*G*), *2G* homozygous (*GG*), or heterozygous (*H*) cell populations incubated in serum free Dulbecco's Modified Eagles Medium (DMEM) containing 0.2% lactalbumin hydrolysate (LH), 10-ng/mL EGF, 10-ng/mL fibroblast growth factor (FGF), or 5-ng/mL IL-1. Each box represents the interquartile range (25th–75th percentile) of the population. The term 1G refers to the sequence 5'-GAA-3' at -1607 bp in the MMP-1 promoter, while 2G refers to the sequence 5'-GGAA-3'. See also Chapter 3. (From Wyatt *et al.*, 2002.)

There are numerous instances where MMPs are involved in development and homeostasis. Studies with chondrocytes demonstrated that these cells undergo a progressive program of proliferation, differentiation, hypertrophy, angiogenesis, and eventually, apoptosis, as they are they replaced with bone at the growth plate; MMPs contribute to this process.[8] Bone is a site of continued tissue remodeling in development, homeostasis, and repair, and bones are continually remodeled throughout life.

MMPs are also included in the development and homeostasis of skeletal and cardiac muscle (Fig. 5.2).[9–11] Secreted and membrane-bound MMP activity has been documented in rat myoblast fusion *in vitro*,[9] suggesting that these enzymes have a role in cell fusion by

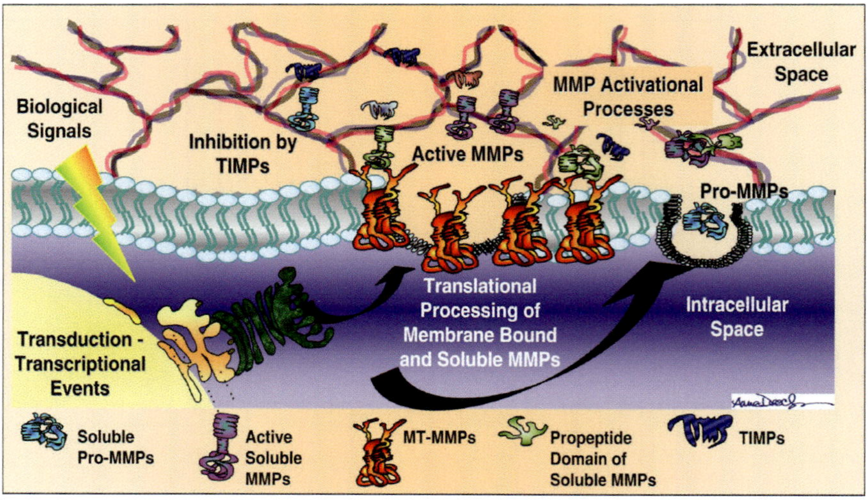

Figure 5.2. A schematic of the current concepts regarding the regulation of MMP activity within the myocardial interstitium. A number of biological signals that include both biochemical and mechanical can affect MMP and TIMP transcriptional activity. While incompletely understood, a number of posttranscriptional steps are required for insertion of the MT-MMPs into the cell membrane as well as release of soluble MMPs. The activation of the soluble MMPs requires proteolytic cleavage of the prodomain, which can occur by other proteases or by the MT-MMPs. This MMP activation step provides for focal proteolytic activity but can quickly amplify as additional MT-MMPs or soluble MMPs are synthesized. The active MMPs (both MT-MMPs and soluble MMPs) are inhibited by the TIMPs. The MMP and TIMP transcriptional and posttranslational events form a dynamic set of interactions that contribute to the overall structure and function of the myocardial intersitium. (From Spinale, 2007.)

favoring cell migration/alignment, which is followed by fusion. In skeletal muscle, latent MMP-2 and MMP-9 are stored the matrix. Then, with stress, such as occurs with myocardial infarction (MI), levels of active enzyme quickly appear, with subsequent modeling and remodeling of cardiac tissues.

Still other experimental systems have shown that MMP-2 and MMP-3 contribute to development of mammary glands (Fig. 5.3).[3,8] Specifically, during puberty, the epithelial ductal network of the gland expands, a process that requires degradation of the basement membrane and extracellular matrix (ECM), restructuring of

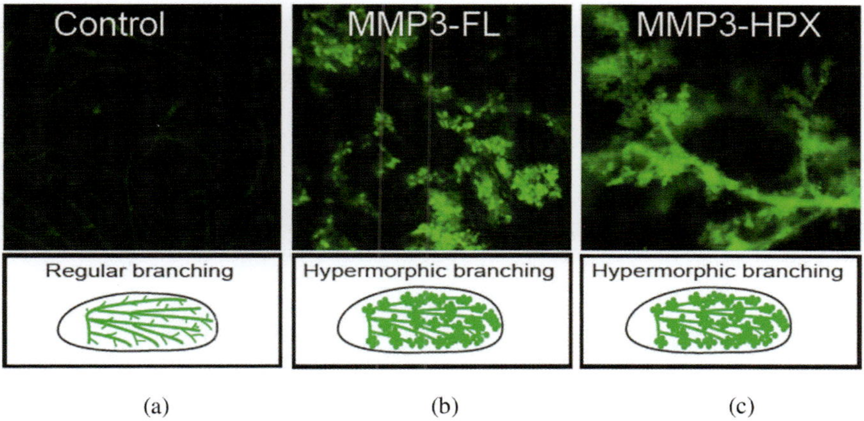

Figure 5.3. The role of MMP-3 in the mammary stem cell niche. MMP-promotes hyperplastic growth in orthotopic transplants of lentivirally transduced mammary epithelial cells. Compared to control transplants (a), overexpression of proteolytically active full-length MMP-3 (b), and MMP3-hemopexin domain (c) both promote a hyperplastic growth phenotype. (From Kessenbrock *et al.*, 2015.)

the vascular network, and epithelial morphogenesis — all activities that implicate MMP involvement. Interestingly, recent work has suggested MMP-3 as contributing to the maintenance of adult epithelial stem cells in the mammary gland.[3] Surprisingly, MMP-3 can mediate effects on stem cell maintenance in the absence of its proteolytic activity, since inactive mutants were sufficient to induce growth of the mammary gland. These studies provide yet another example of complex functions of MMPs.

5.3 "Trials and Tribulations" with MMPIs

It was the studies with broad-spectrum MMPIs targeted at the enzymatically active site(s) that revealed unexpected findings about the subtle and ongoing actions of MMPs in homeostasis. When these inhibitors were first introduced in the late 1980s, there was great excitement about their efficacy as therapeutic agents in conditions where their over expression was rampant (Table 5.1).[1,12] In 1992, British Biotech introduced Batimastat (BB-94). BB-94 is a

Table 5.1. Phase III clinical trials with MPIs.

MPI	Cancer	Stage	Treatment	Result
Marimastat (BB-2516)	Pancreatic [35]	II, III, IV; unresectable	5, 10, 25 mg vs. gemcitabine	No significant difference in overall survival; in subset analysis, 25 mg had 1-year survival rate similar to gemcitabine
	Pancreatic [17]	II, III, IV; unresectable	Gemcitabine with 10 mg or placebo	No survival benefit
	Pancreatic [36]	I, II, III; resectable	20 mg versus placebo	Result expected December 2002
	Gastric [36, 37]	Advanced unresectable	10 mg versus placebo	No or very modest survival benefit ($P = 0.07$); in subset analysis, significant survival benefit in patients who received prior Rx ($P = 0.045$)
	Glioblastoma [17, 56]	Unresectable	10 mg versus placebo	No survival benefit
	Small-cell lung [57]	Any; PR or CR after first Rx	10 mg versus placebo	No survival benefit
	Non–small-cell lung [36]	IIIA or IIIB Advanced second Rx	10 mg versus placebo carboplatin with 10 mg or placebo	No survival benefit No difference in response

Prinomastat (AG3340)	Non–small-cell lung [44]	IIIBT4 or IV	Carboplatin + paclitaxel with prinomastat or placebo	No survival benefit
	Non-small-cell lung [44]	IIIBT4 or IV	Cisplatin + gemcitabine with prinomastat or placebo	Terminated early because of lack of efficacy
	Prostate [44]†	Metastatic, hormone refractory	Mitoxantrone + prednisone with prinomastat or placebo	No difference in symptomatic progression
Tanomastat (BAY 12-9566)	Small-cell lung [38]	Extensive; PR or CR after first Rx	Tanomastat versus placebo	Terminated prematurely because tanomastat-treated patients showed poorer survival than placebo-treated patients
	Pancreatic [39]	Metastatiic; unresectable; no prior Rx	Gemcitabine versus tanomastat	Terminated prematurely because tanomastat-treated patients showed poorer survival than gemcitabine-treated patients
BMS-275291	Non–small-cell lung [43]	IIIB or IV	Carboplatin + paclitaxel with BMS-275291 or placebo	Currently recruiting patients
Neovastat	Renal cell carcinoma [46]	IV	Neovastat vs. placebo	Currently recruiting patients

Source: From Coussens (2009). The references listed in this table relate to the original source.

Endpoint for all studies was survival except: *Endpoint was response. †Endpoint was time to symptomatic progressive disease. Abbreviations: PR, partial response; CR, complete response; Rx, treatment.

broad-spectrum hydroxamic-acid memetic based on the structure of collagen. It was the first MMPI to be tested in phase I trials,[1] but the drug was plagued with poor solubility, complicating issues of drug delivery and interpretation of clinical data. Although the poor solubility was overcome with intrapleural or intraperitoneal injections, which created a depot of drug with sustained plasma levels, the drug remained problematic.

Consequently, clinical trials with Batimastat were soon replaced with Marimastat, an orally available cousin. Unfortunately, despite some initial therapeutic success, prolonged treatment resulted in debilitating joint pain.[1,12] Eventually, these crippling side effects were attributed to widespread inhibition of MMP enzymatic activities, a concept that raised the intriguing possibility that low levels of these enzymes may actually be beneficial in maintaining normal physiology. However, there was uncertainty about (1) which MMPs were involved in maintaining normal connective tissue functions and (2) which MMPs needed to be inhibited for therapeutic efficacy in diseases such as cancer and arthritis. Finally, it became clear that the issue was quantitative: it was the level of MMPs that mattered, rather than which one was expressed.

The focus then shifted to developing inhibitors that selectively targeted particular MMPs, and by 2000, nearly all of the large pharmaceutical firms had MMPIs either in the pipeline or in early clinical trials. Prinomastat (AG-3340), another hydroxamic-acid derivative, is relatively selective for the gelatinase, MMP-2, as compared with collagenase, MMP-1. Nonetheless, it also showed disappointing side effects, similar to Marimastat, suggesting either that Prinomastat was not selective and/or that MMP-2 had a substantial role in homeostasis so that blocking its activity resulted in the musculoskeletal side-effects. On the other hand, Bayer's BAY12-9566, a compound with a similar selectivity to Prinomastat did not elicit musculoskeletal side effects, but failed to show clinical efficacy. Frustration over the "tribulations" with MMPIs and their failed clinical trials continued to grow. In fact, there are some reports that patients with this MMPI actually fared worse than their placebo-treated counterparts,[1,12] again suggesting that the drug blocked the enzymatic activities of MMPs that were physiologically beneficial in homeostasis of our connective tissues. Thus, by early 2002, interest in MMPIs was abandoned due to the abysmal failure of these clinical trials.[1,12] However, these failed trials left important clues about the essential roles of MMPs in normal physiology and ECM remodeling.

5.4 MMP Knockout Mice

Genetic studies with the selective deletion of an MMP (MMP "knockout" mice) provided additional information about the function of MMPs in development and homeostasis.[9,13] Indeed, the most detailed data on the importance and roles of MMPs in development and in normal physiology have come from analysis of mice with a selective deletion(s) of MMPs. These studies confirmed the suspicions about the redundant roles of MMPs and expanded our insights into their contributions to disease pathologies. Perhaps even more importantly, they began to more clearly define the beneficial roles of MMPs in normal physiology. So far, 16 MMP knockout mutants have been analyzed, along with four double knockouts (Table 5.2).[13]

Mice with a single MMP deletion are surprisingly devoid of a severe phenotype (Table 5.2).[1,13] Most mice survive well, and the absence of the MMP becomes apparent only in response to an environmental challenge, such as infection or wound healing. These relatively benign phenotypes support the idea of functional redundancy among MMPs as far as their substrates, with the results often being enzymatic compensation. Nonetheless, despite this redundancy, the absence of a particular MMP has revealed some (1) abnormalities in bone growth and remodeling, (2) defects in angiogenesis and vascular development, especially in pathological conditions, and (3) dysregulation of some immune responses.

Although the functional homologue of human MMP-1 in rats and mice is MMP-13, recent studies have identified murine MMP-1, Mmp1a. With only limited sequence identity to human MMP-1 and transient period of expression during embryogenesis, murine Mmp1a was thought to be a nonfunctional relic (Fig. 5.4).[1,14] Recently, however, its expression was shown to be substantially upregulated in

Table 5.2. Phenotypes of MMP knockout mice.

MMP	Phenotype
MMP-2	No overt phenotype, reduced body size
	Reduced neovasculation and tumour progression
	Decreased primary ductal invasion in the mammary gland
	Decreased bone mineralisation, joint erosion, defects in osteoblast and osteoclast growth
MMP-3	No overt phenotype
	Impaired contact dermatitis, impaired onset of T-cell proliferation
	Defect in wound contraction
MMP-7	Altered secondary branching morphogenesis in the mammary gland
	No overt phenotype
	Impaired tracheal wound repair
	Innate immunity defects
	Defective prostate involution
MMP-8	More susceptible to develop skin cancer, impaired wound healing in skin
	Altered inflammatory response in wounds, delay of neutrophil infiltration
	Altered TGF-β signalling
MMP-9	Impaired primary angiogenesis in bone growth plates
	Resistant to bullous pemphigoid
	Contact dermatitis: delayed resolution
	Impaired vascular remodelling
	Delayed healing of bone fractures
MMP-10	No overt phenotype
	Pathological induced phenotype: pulmonary inflammation and mortality

MMP-11	No overt phenotype Decreased chemical-induced mutagenesis
MMP-12	No overt phenotype Impaired macrophage proteolysis Resistant to cigarette smoke-induced emphysema
MMP-13	No overt phenotype Induction of MMP-8 expression in *MMP13-/-* wounds, no different in wound healing Bone remodelling defects
MMP-14 (MT1-MMP)	Premature death; skeletal defects and dwarfism Normal at birth but develop multiple abnormalities (defect in remodelling of the connective tissue, increased bone resorption and defective secondary ossification centres) and die by 3–12 weeks Angiogenesis defect and defects in lung and submandibular gland
MMP-16 (MT3-MMP)	Growth retardation
MMP-17 (MT4-MMP)	No overt phenotype
MMP-19	Obesity and decreased skin carcinogenesis
MMP-20	Defects in tooth enamel
MMP-24	Abnormal response to sciatic nerve injury
MMP-28	No overt phenotype; elevated macrophage recruitment in lung
MMP-2/MMP-9	Impaired tumour invasion and angiogenesis
MMP-2/MMP-14	Die immediately after birth with respiratory failure, abnormal vessels and immature muscle fibres
MMP-9/MMP-13	Shortened bones
MMP-14/MMP-16	Die one day after birth due to cleft palate

Source: From Löffek *et al.* (2011).
TGF: transforming growth factor.

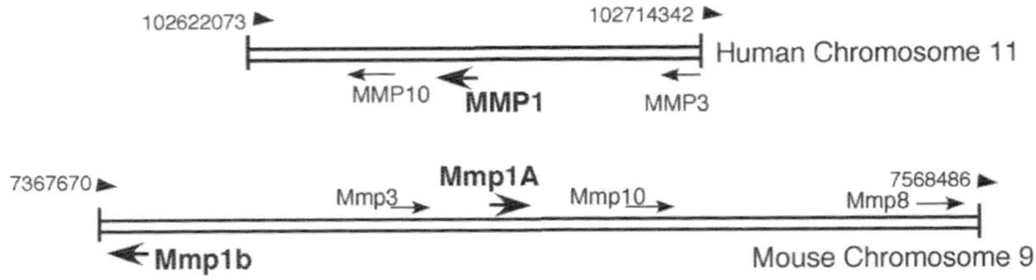

(a)

	Mus musculus (1A)	*Mus musculus (1B)*	*Rattus norvegicus (1A)*	*Rattus norvegicus (1B)*	*Oryctolagus cuniculus*	*Sus scrofa*	*Canis lupus*	*Bos taurus*	*Equus caballus*	*Pan troglodytes*
Homo sapien	57.3	55.1	56.2	54.3	86.1	84.0	79.9	83.8	85.5	98.7
Mus musculus (1A)		81.6	88.1	76.3	55.6	55.8	56.5	56.5	58.0	57.1
Mus musculus (1B)			76.5	81.5	54.0	54.0	51.8	53.8	54.9	54.4
Rattus norvegicus (1A)				77.6	55.4	55.8	56.5	55.2	58.2	55.6
Rattus norvegicus (1B)					52.6	52.4	53.0	52.6	54.3	53.5
Oryctolagus cuniculus						82.7	78.8	83.5	84.8	85.5
Sus scrofa							80.3	86.1	86.6	83.5
Canis lupus								80.8	83.1	79.5
Bos taurus									85.9	83.5
Equus caballus										85.0

(b)

Figure 5.4. Homology of human MMP-1 and murine Mmp-1a. (a) Chromosomal location of MMP1 (upper) and Mmp1a (lower) in the MMP rich locus of humans and mice, respectively. (b) Percent similarity of mammalian MMP1 homologue protein sequences. (From Foley and Kuliopulos, 2014.)

inflammation and in some mouse models of cancer.[14] Mmp1a protein is inherently unstable,[14] perhaps accounting for its previously unrecognized roles in murine physiology and pathology. However, the creation of an Mmp1a deficient mouse has revealed that this enzyme contributes to the progression of lung cancer, since ablation of this gene resulted in smaller tumors and impaired angiogenesis. The defect in tumorigenesis in the null mice was mediated by the absence of Mmp1a/PAR-1 signaling, analogous to the activation of this signal transduction pathway described for human MMP-1.[14,15] Further, co-implanting lung tumor cells with fibroblasts from wild-type mice rescued the growth of tumors in the Mmp1a –/– mice, indicating an important role for stromal Mmp1a.

Abnormalities in skeletal remodeling and/or delays in growth are apparent in mice lacking MMP-2, MMP-9, and MMP-13. Mice without MMP-2 (gelatinase A) have growth delays and skull deformities, due to intramembranous ossification defects,[13] while mice lacking MMP-9 (gelatinase B) show retarded development and growth of long bones, due to improper vascular invasion in skeletal growth plates. These findings implicate MMP-9 in vascular remodeling (Table 5.2) and in modifying the ECM.[13] However, by adulthood, this phenotype is mild, with adults showing only about 10% shortening in the long bones, again illustrating redundancy and compensation by other MMPs. The absence of the interstitial collagenase, MMP-13, gives rise to aberrant development of the skeletal growth plate and the delayed exit of chondrocytes from the growth plate, demonstrating the importance of collagenolysis in bone maturation. The absence of collagenolytic activity is also linked to the phenotype of mice lacking MMP-16 (MT3-MMP).[13] Here, the growth of mice is impaired because of decreased viability of mesenchymal cells in skeletal tissues. This results from the inability of the cells to degrade collagen fibers, which interferes with their ability to proliferate and to migrate.[13] These examples illustrate the crucial role(s) of MMPs in modulating the microenvironment surrounding cells so that they can carry out the functions required during normal development.

Among all single MMP knockout models, only MMP-14 (MT1-MMP) deficient mice exhibit a profound and severe phenotype, even in the absence of an environmental challenge (Table 5.2; Fig. 5.5).[16] These mice live only a few weeks and display multiple skeletal defects including craniofacial dysmorphism, arthritis, osteopenia, dwarfism, and fibrosis of soft tissues, and once again, they illustrate how the activity of MMPs influences the cellular microenvironment. All of these defects result from the lack of collagenolytic activity that is necessary for modeling of skeletal and surrounding supporting connective tissues. However, in this case, MT1-MMP is absolutely required for connective tissue metabolism during development, and compensatory mechanisms are not present.

MMPs have major functions in the immune system and innate immunity, and MMP-7 (matrilysin) is a principle player.

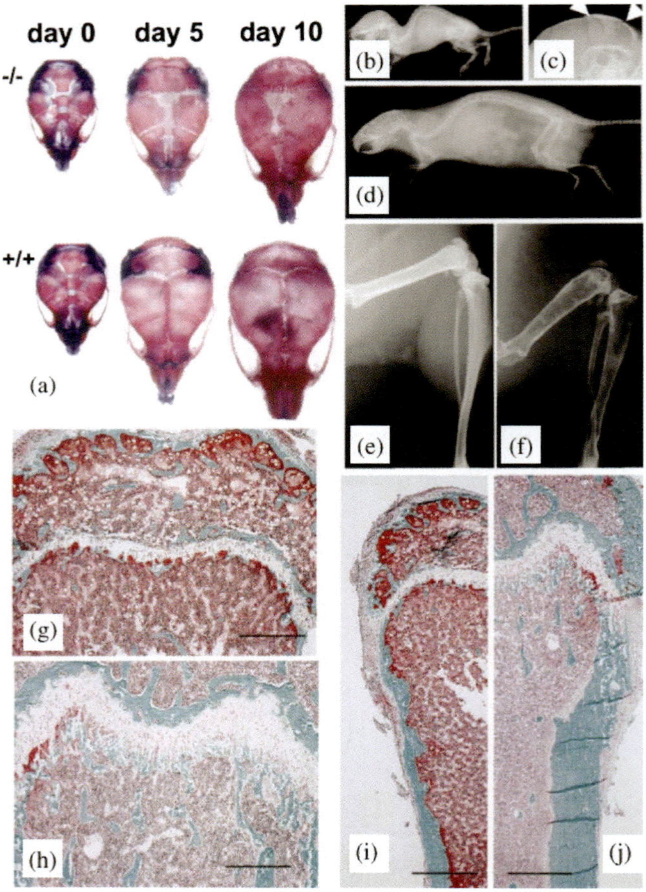

Figure 5.5. Bone development in MT1-MMP-deficient mice (a) Alizarin red/alcian blue staining of skulls of MT1-MMP-deficient (–/–) and control (+/+) mice demonstrating larger fontanelles and gradually increasing cranial dysmorphism of mutant mice (3× magnification). (b and d) Lateral X-ray images of 80-day-old MT1-MMP-deficient (b) and wild-type littermate (d), (0.75× magnification). (c) 3′ magnification of cranial vault from the animal in (b). Note the wide sutures and the displacement of the interparietal bone (arrowheads). (e and f) X-ray images of hind limb bones from 80-day-old MT1-MMP-deficient (f) (3× magnification) and wild-type littermate (e) (2′ magnification). Note the severe osteopenia of the mutant mouse. (g–j) Histology of the femora of 60-day-old MT1-MMP-deficient (g and i) and wild-type (h and j) mice. Note the severe reduction of trabecular bone in the spongiosa and the thinner cortex in the mutant animal (g–h). Goldner's stain. Bars = 150 μm (g and h); 400 mm (i and j). (From Holmbeck *et al.*, 1999.)

MMP-7 null mice are susceptible to bacterial intestinal infection, partly because they cannot release the endogenous antibiotic peptide, α-defensin, from intestinal epithelia.[13] MMP-7 in the lung is also required for the chemotactic attraction of neutrophils to damaged lung epithelium. In wild-type mice, MMP-7 proteolytically releases syndecan-1, which sequesters a number of cytokines that form a gradient for attracting neutrophils. In the absence of MMP-7, syndecan is not released, no gradient is formed, and no neutrophils are recruited.

Since MMPs have shown redundant proteolytic activity against the same substrates *in vitro*, it is possible that there is a similar redundancy *in vivo*.[1,13,17,18] Indeed, knockout mice with two MMPs deleted display severe phenotypes, suggesting that any compensatory mechanism has been eliminated. A dramatic example is the MMP-2/MT1-MMP double null mouse, which dies immediately after birth from respiratory failure, abnormal vessel formation, and immature muscle fibers.[13] The pathology arising from deleting two MMPs provides strong evidence for redundant functions of MMP *in vivo* during embryonic development.

Double MMP knockouts have also emphasized the physiological role of MT3-MMP (MMP-16) and MT1-MMP (MMP-14). Mice that are double null for these MMPs exhibit lethal embryonic defects in development of the palate and in bone formation.[13] Once more, these defects are due to absence of collagen degradation, degradation that is required for mesenchymal cells to remodel a collagen-rich matrix and then to proliferate during embryogenesis.

5.5 MMPs as "Good Guys"

It is clear that the total absence of an MMP may have some detrimental effects on normal development and homeostasis. This strongly suggests that some level of MMP activity is essential, and perhaps even beneficial and/or protective.[1,18,19] For example, MMP-3 knockout mice developed collagen-induced arthritis just as frequently, if not more so, than wild-type counterparts.[1] There is a surprising finding, since MMP-3 levels are aberrantly high in arthritis. In another example, MMP-3 null mice develop squamous cell carcinomas, and the tumors were more undifferentiated compared to controls, with a concomitant increase in lung metastases.[18] Further, a protective role for MMP-3 in tumor progression is suggested by the finding that the tumors in these MMP-3–/– showed reduced leukocyte infiltrate.[18] However, high MMP-3 expression can also be protumorigenic,[18] implying that the level of MMP-3 expression is critically important: a total absence can result in one type of pathology, while too much results in another. Perhaps there is a "Goldi-Locks" amount of MMPs that is "just right".

Deletion of MMP-8 (collagenase-2; neutrophil collagenase) has provided the most striking evidence of a protective/beneficial effect of an MMP and, consequently, MMP-8 has been designated as an "anti-target" for cancer therapy (Table 5.3).[1,18,19] Male mice that are null for MMP-8 display increased susceptibility to skin tumors. A hormonal effect is implicated since ovariectomizing female mice increases tumor incidence to that seen with male mice.

Other studies with breast cancer cell lines have shown that overexpression of MMP-8 was associated with a non metastatic cell line, and that this protection was lost if MMP-8 expression was knocked down. Additional studies with a mouse model of melanoma demonstrated that overexpression of MMP-8 resulted in far fewer metastatic lesions. Interestingly, these findings were supported by a report that MMP-8 is often mutated in human melanomas. Wild-type MMP-8 can inhibit growth of human melanoma cells in soft agar *in vitro* and reduce tumor formation *in vivo*. In contrast, mutants of MMP-8 cannot mediate these activities, suggesting that MMP-8 can block melanoma progression.[19] Additional data implicate MMP-8 in modulating protective immune responses, probably by helping to recruit neutrophils and increases in adhesion to type I collagen and laminin.[19] Therefore, as evidence of its protective role in cancer continues to accrue, MMP-8 seems to function as a tumor suppressor gene.[1,18,19]

A protective role for MMP-9 in asthma has also been suggested.[18] Overexpression of MMP-9 in lung tissues has been associated with the pathology of asthma. However, compared to normal

Table 5.3. Protective roles of MMP-8.

MMP8	References	Study type	Cancer	Main findings
	[34]	*In vivo*	Skin cancer (chemical induced)	Pro-tumorigenic in male KO mice Protective effect restored by bone marrow transplantation from WT mice
	[35]	*In vivo*	Breast cancer (MDA-MB-435 cell line)	Elevated expression in non-metastatic-derived cell line Increased migration through Matrigel in absence of MMP8
	[36]	*In vivo*	Breast cancer (MDA-MB-435 cell line)	Pro-metastatic with ribozyme knockdown
	[37]	*In vivo*	Melanoma (Mel-STR cell line)	Inhibition of cell proliferation
		In vivo	Melanoma (Mel-STR cell line)	Inhibition of tumour growth
	[38]	*In vivo*	Tongue squamous carcinoma (chemical induced)	Pro-tumorigenic in female KO mice
		Human studies	Tongue squamous carcinoma	Prolonged OS
	[39]	Human studies	Breast cancer	Plasma levels positively associated with lymph node metastasis, negatively associated with distant metastasis

(*Continued*)

Table 5.3. (*Continued*)

MMP8	References	Study type	Cancer	Main findings
	[40]	Human studies	Breast cancer	SNP associated with reduced lymph node metastasis rs11225395 SNP confers better prognosis (DFS, OS)
	[41]	Human studies	Lung cancer	SNP associated with decreased risk of lung cancer
	[44]	*In vivo*	Melanoma (B16F10 cell line)	Inhibition of invasion and transendothelial migration Increased cell adhesion to collagen-1, laminin-1 No effect on cell proliferation
		In vivo	Melanoma(B16F10 cell line)	Anti-metastatic No effect on tumour growth
		Human studies	Breast cancer	Inversely associated with lymph node metastasis

Source: From Decock *et al.* (2011). The references listed in this table relate to the original source.
DFS: disease-free survival: KO: knockout and OS: overall survival.

mice, if MMP-9 null mice are challenged with experimental asthma, the mice have trouble resolving the inflammatory response.[18] These findings strongly imply that too little or too much of an MMP can be associated with disease symptoms, and underscore the concept that low basal levels are essential for maintaining homeostasis. Even subtle perturbations can have "bad' consequences.

In summary, investigations with MMP knockout mice have definitively revealed essential roles for these enzymes in normal embryonic development. In addition, experiments with double knockouts strongly support the concept of functional enzymatic redundancy, since single knockout mice often display relatively benign phenotypes, with severe phenotypes only apparent when redundancy is no longer operational, that is, double nulls. Further, double knockout mice have demonstrated that the complete absence of an MMP, along with its redundant counterpart, can be deleterious, indicating that MMPs have critical roles in homeostasis. Many of these roles appear to involve control of inflammation and innate immune responses. Finally, not only are MMPs crucial in normal physiology, their regulated and tightly controlled expression is actually beneficial and protective. In the end, it may depend on the "correct" level of expression of MMPs at the appropriate time and place during embryonic development and in the course of normal physiology in adult life. What represents the "correct" level of a particular MMP, and where and when this expression is needed, may vary according to the spatial and temporal requirements of our body.

5.6 MMPs and Tissue Repair

The recovery of normal tissue architecture and function after disruption by injury or infection is an exquisite process, with multiple carefully regulated steps.[4,13,20] Both secreted and membrane-bound MMPs are critical players as they facilitate cell migration and neovascularization. The normal low basal levels of MMP expression are challenged with a wound or an infection, and this challenge sets in motion a series of controlled cellular and molecular events. The end result is wound closure and healing, with the acute response silenced, and levels of MMP expression returning to basal levels. In contrast, in cancers and some other diseases, there is an excessive and aberrant MMP response, leading to "wounds that do not heal".[4] Therefore, an examination of wound healing is an excellent model for demonstrating the superbly regulated steps involved in wound healing and for illustrating how this process goes awry in disease.

Tissue injury induces expression of most MMPs in most of the cells that are either already present or that are recruited to the site of injury: platelets, mesenchymal, epithelial, and immune cells (Fig. 5.6).[4,13,20] The attraction of platelets is among the earliest events, resulting in the formation of a fibrin/fibronectin clot. This clot provides a rudimentary ECM and activates epithelial cells at the edge of the wound to begin the process of closing the denuded tissue. The wounded epithelial cells, and other cells present in the adjacent microenvironment (e.g., leukocytes, endothelial cells, and fibroblasts), release chemokines and cytokines that attract cells of the innate immune response (neutrophils and macrophages), which are then followed by cells of the adaptive immune response (regulatory T cells). Simultaneously, the fibroblasts are activated to become myofibroblasts and they begin to deposit ECM. The temporal and spatial expression of MMPs is one important component of these activities (see Fig. 4.4; Fig. 5.7).[13,20]

Since wounding of the skin is very common, it is an appropriate model for study, even though epithelial and mucosal lining cells of the gastrointestinal (GI) tract, lungs, kidney tubules are also often subjected to injury.[13,20] During wound healing, epithelial cells in human skin move across the bed of the wound, which is a substratum in only two dimensions but where MMP-1 facilitates reepithelialization.[13,20] The keratinocytes have a strong binding affinity for $\alpha 2\beta 1$ integrin, which prevents them from migrating across the wound. However, keratinocytes at the leading edge of the wound are induced to express high levels of MMP-1 when their $\alpha 2\beta 1$ integrin binds to the dermal collagen type I. The increase of MMP-1 destroys the collagen matrix, reducing the affinity of the integrin-collagen binding, and thus permits the migration of the keratinocytes. The reduction in binding affinity also reduces MMP-1 expression.

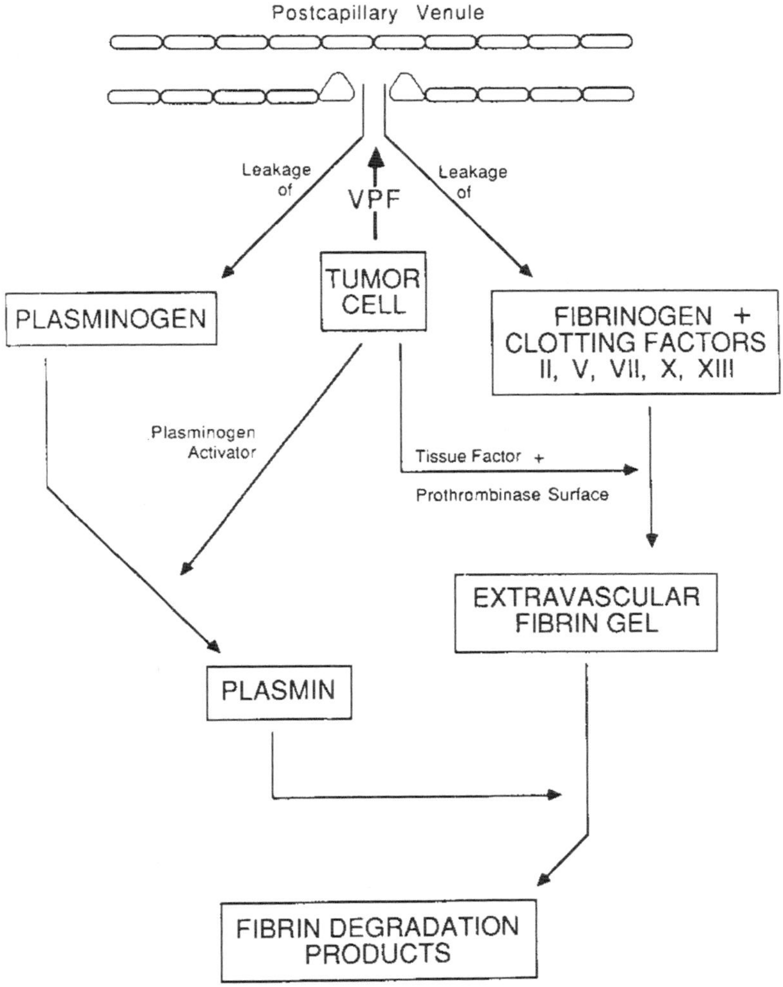

Figure 5.6. Tumor-host cell interactions that regualte fibrinogen influx, extravascular coagulation, fibrin deposition, and fibrin turnover in solid tumors. VPF, vascular permeabilty factor. (From Dvorak, 1986.)

Importantly, there is an absolute requirement for MMP-1 in mediating migration on collagen since MMP-1 blocking antibodies or mutant collagen that is resistant to degradation completely prevents migration, even though other collagenolytic enzymes such as MMP-13 and MT1-MMP may also be produced by keratinocytes at the wound edge.[20] This process of (1) inducing MMP-1 by high affinity binding to α2β1 integrin to collagen, (2) reducing the affinity to collagen, (3) allowing cell migration, and then (4) lowering MMP-1 expression continues as the keratinocytes travel over the surface of the open wound.[13,20] It has been suggest that "by repeatedly establishing tight contacts, rapidly loosening this hold by the action of MMP-1, and reestablishing new tight contacts, keratinocytes use native type I collagen as a 'compass' to guide repair".[20]

Another critically important aspect of wound healing is new vessel formation, which involves movement and proliferation of endothelial cells.[13] Neoangiogenesis is induced by fibrin and fibronectin present in "granulation tissue", an old term that describes the granular appearance of new connective tissue laid down at the site of the wound.[4,13] MMPs cause the release of proangiogenic factors that bind to the newly formed ECM. Specifically, MMP-2 and MMP-9 release vascular endothelial growth factor (VEGF) and tumor necrosis factor-alpha (TNF-α), while the complex of MT1-MMP/TIMP-2 expressed on the surface of activated endothelial cells activates pro-MMP-2 to its catalytically active form. In addition, MT1-MMP colocalizes with αvβ3 integrin at the intercellular junctions of endothelial cells, where it facilitates endothelial cells migration and adhesion (see Fig. 4.4).[13]

MMP-7 also has significant functions in wound repair, especially in the GI tract and lung (Fig. 5.7).[13,20] Along with recruiting neutrophils as part of an innate immune response, MMP-7 is required for reepithelialization in colon and lung injury, a process that is totally suppressed in MMP-7 null mice. The mechanism by which MMP-7 facilitates wound healing appears to involve its ability to shed syndecan-1, a transmembrane heparin sulfate proteoglycan, from the surface of epithelial surface. This shedding event lessens the binding of α2β1 to matrix substrates, thereby relieving restraints on cell migration. This phenomenon is similar to that seen with the keratinocytes and MMP-1 in the skin.[13,20]

Lastly, it is important to consider the role of MMPs in chronic wounds, especially in vascular disease of the legs and feet. In these conditions, successful healing may be difficult or severely

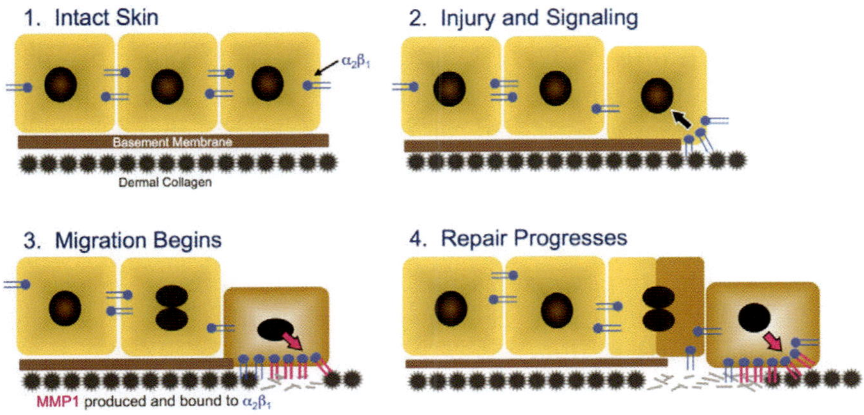

Figure 5.7. Summary of proposed mechanisms of how MMP-1 and MMP-7 facilitate reepithelialization in different tissues. (From Rohani and Parks, 2015.)

compromised when the carefully choreographed events of normal wound healing are overrun by the mediators of chronic venous disease (CVD).[21–23] CVDs can be very debilitating, affecting millions of people and resulting in varicose veins and/or in venous ulcers, that when severe, extend to changes in the skin. Although genetic and environmental factors may predispose certain individuals to varicose veins and CVD, their underlying basis is chronic inflammation in the venous circulation. As a result, pressure within the vein does not fall during exercise (normal condition), but rather rises, indicating that the deep veins are occluded. The endothelium then senses changes in shear stress. This leads to the activation of macrophages, monocytes, and T-lymphocytes, and release of their inflammatory mediators: chemokines, cytokine, growth factors, and MMPs. These mediators trigger chronic stimulation of signal transduction pathways and changes in gene expression, eventually causing changes in the structure of the venous wall and valves, dilation of the veins varicosity, and advanced forms of CVD (Fig. 5.8).

Chronic venous ulcers are characterized by abnormally high levels of proteinases, especially MMPs, in the tissue microenvironment. Despite the fact that MMPs are needed for normal healing to occur, the higher levels of MMPs in exudates from chronic wounds compared to those of acute wounds impair the healing process. This is probably because dysregulated expression of some MMPs disrupts the tightly regulated sequence of events that normally occur in wound closure. Nonetheless, MMP-1 is the primary collagenase involved in successful wound closure, and its ability to degrade type I collagen in the dermis is essential for keratinocyte migration and re-epidermization. Therefore, higher levels of MMP-1 in chronic wounds can be associated with a better prognosis for satisfactory healing.

High levels of MMP-8 and MMP-9 are also found in chronic wounds. These MMPs decrease in chronic wounds that eventually heal, but not in wounds that heal poorly. Since these MMPs are products of inflammatory cells, their decrease suggests a decline in inflammation, thereby facilitating successful healing. Concomitantly, the higher levels of MMP-9 seen in poor healers lend further support to the concept that MMPs associated with inflammatory cells impede healing. Consequently, it has been suggested that they may be appropriate therapeutic targets designed to block their activities.[22] In addition, it has been suggested that future studies that focus on changes in cellular metabolism in CVDs and on the potential role of microRNAs (mi-RNAs) as mediators of pathology may provide insights for developing new treatments (Fig. 5.8).[23]

5.7 Conclusions

Given the compelling data that MMPs are essential and even beneficial for normal physiology and homeostasis, understanding their various roles in disease pathologies becomes a challenging task. It is easy to dismiss this as simple overexpression and dysregulation. However, this is naïve since ultimately, the problem is the mechanism(s) controlling the activation of signal transduction pathways and regulation of gene expression and enzymatic activity that put MMPs in the "on" position. This assumes, as is probably mostly correct, that the rampant overexpression of a particular MMP(s) contributes to disease pathology. However, it is important to remember that in a few instances, the absence of an MMP (as in

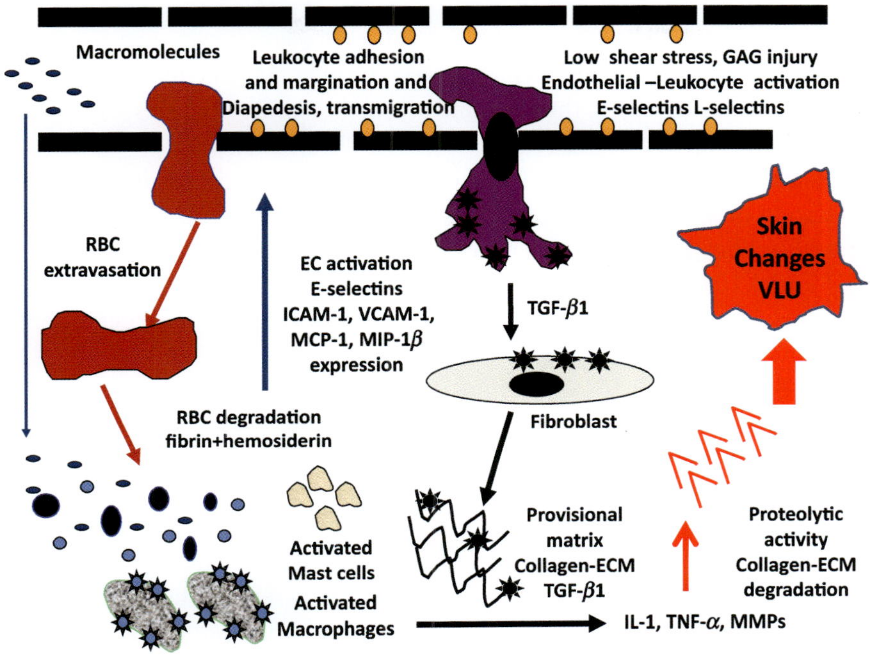

Figure 5.8. Schematic diagram of endothelial cell activation and leukocyte activation. Expression of several chemoattractants, cytokines, and MMP. The consequences of unabated inflammation and leukocyte activation are the expression of several cytokines and MMPs with degradation of collagen and the ECM leading to skin changes and venous leg ulcer. (From Raffetto and Mannello, 2014.)

MMP-8 in cancer and MMP-9 in asthma) may exacerbate pathology. Given what we know about the orchestrated program for wound healing, it is not difficult to see how a cancer or other disease with constant high levels of MMP expression could have serious consequences. As Dvorak stated: Tumors "mimic wounds by depositing fibrin-fibronectin gel. Such gels, in tumors as at sites of local injury, signal the host to marshal the wound-healing response. This response is stereotyped and similar in both tumors and wounds. In tumors, however, the fibrin-fibronectin matrix signal that evokes the wound-healing response is not self-limited.... Thus tumors appear to the host in the guise of wounds, or more correctly, of an unending series of wounds that continually initiate healing but never heal completely".[4]

References

1. Brinckerhoff, CE and LM Matrisian (2002). Matrix metalloproteinases: a tail of a frog that became a prince. *Nature Reviews Molecular Cell Biology*, 3, 207–214.
2. Gaffney, J, I Solomonov, E Zehorai and I Sagi (2015). Multilevel regulation of matrix metalloproteinases in tissue homeostasis indicates their molecular specificity in vivo. *Matrix Biology*, 44–46, 191–199. doi:10.1016/j.matbio.2015.01.012.
3. Kessenbrock, K, C-Y Wang and Z Werb (2015). Matrix metalloproteinases in stem cell regulation and cancer. *Matrix Biology*, 44–46, 184–190. doi:10.1016/j.matbio.2015.01.022.
4. Dvorak HF (1986). Tumors: wounds that do not heal. Similarities between tumor stroma generation and wound healing. *New England Journal of Medicine*, 315, 1650–1659.
5. Mizel, SB, JM Dayer, SM Krane and SE Mergenhagen (1981). Stimulation of rheumatoid synovial cell collagenase and prostaglandin production by partially purified lymphocyte-activating factor (interleukin 1). *Proceedings of the National Academy of Sciences of the U.S.A.*, 78, 2474–2477.
6. Wyatt, CA, CI Coon, JJ Gibson and CE Brinckerhoff (2002). Potential for the 2G single nucleotide polymorphism in the promoter of matrix metalloproteinase to enhance gene expression in normal stromal cells advances in brief potential for the 2G single nucleotide polymorphism in the promoter of matrix metalloprote. *Cancer Research*, 7200–7202.
7. Malemud, CJ (2006). Matrix metalloproteinases (MMPs) in health and disease: an overview. *Frontiers in Bioscience*, 11, 1696–1701.
8. Page-McCaw, A, AJ Ewald and Z Werb (2007). Matrix metalloproteinases and the regulation of tissue remodelling. *Nature Reviews Molecular Cell Biology*, 8, 221–233.
9. Alameddine, HS (2012). Matrix metalloproteinases in skeletal muscles:friends or foes? *Neurobiology Disease*, 48, 508–518.
10. Kwak, H-B (2013). Aging, exercise, and extracellular matrix in the heart. *Journal of Exercise Rehabiltation*, 9, 338–347.
11. Spinale, FG (2007). Myocardial matrix remodeling and the matrix metalloproteinases: influence on cardiac form and function. *Physiological Reviews*, 87, 1285–1342. doi:10.1152/physrev.00012.2007.
12. Coussens, LM (2009). Matrix metalloproteinase inhibitors and cancer: trials and tribulations. *Science*, 2387, 2387–2393.

13. Löffek, S, Schilling, O and C-W Franzke (2011). Series 'matrix metalloproteinases in lung health and disease': biological role of matrix metalloproteinases: a critical balance. *European Respiratory Journal*, 38, 191–208.
14. Foley, CJ and A Kuliopulos (2014). Mouse matrix metalloprotease-1a (Mmp1a) gives new insight into MMP function. *Journal of Cellular Physiology*, 1875–1880. doi:10.1002/jcp.24650.
15. Blackburn, JS, I Liu, CI Coon and CE Brinckerhoff (2009). A matrix metalloproteinase-1/protease activated receptor-1 signaling axis promotes melanoma invasion and metastasis. *Oncogene*, 28, 4237–4248.
16. Holmbeck, K *et al.* (1999). MT1-MMP-deficient mice develop dwarfism, osteopenia, arthritis, and connective tissue disease due to inadequate collagen turnover. *Cell*, 99, 81–92.
17. Hadler-Olsen, E, B Fadnes, I Sylte, L Uhlin-Hansen and JO Winberg (2011). Regulation of matrix metalloproteinase activity in health and disease. *FEBS Journal*, 278, 28–45.
18. Vincenti, MP and CE Brinckerhoff (2007). Signal transduction and cell-type specific regulation of matrix metalloproteinase gene expression: can MMPs be good for you? *Journal of Cellular Physiology*, 213, 355–364. doi:10.1002/jcp.21208.
19. Decock, J, S Thirkettle, L Wagstaff and DR Edwards (2011). Matrix metalloproteinases: protective roles in cancer. *Journal of Cellular and Molecular Medicine*, 15, 1254–1265.
20. Rohani, MG and WC Parks (2015). Matrix remodeling by MMPs during wound repair. *Matrix Biology*, 44–46, 113–121. doi:10.1016/j.matbio.2015.03.002.
21. Mannello, F, D Ligi and JD Raffetto (2014). Glycosaminoglycan sulodexide modulates inflammatory pathways in chronic venous disease. *International Angiology*, 33, 236–242.
22. Muller, M *et al.* (2008). Matrix metalloproteinases and diabetic foot ulcers: the ratio of MMP-1 to TIMP-1 is a predictor of wound healing. *Diabetic Medicine*, 25, 419–426.
23. Raffetto, JD and FM Mannello (2014). Pathophysiology of chronic venous disease. *Journal of Angiology*, 33, 212–221.

6. Matrix Metalloproteinases (MMPs) and Cytokines in Rheumatology

Jean-Michel Dayer*, Samantha J. Williamson†
Adam P. Croft† Christopher D. Buckley†
and Carlo Chizzolini‡

**Medical School, University of Geneva, Geneva, Switzerland*
†Institute of Inflammation and Ageing,
College of Medical and Dental Sciences,
University of Birmingham, Birmingham, United Kingdom
‡Division of Immunology and Allergy,
University Hospital and Medical School,
University of Geneva, Geneva, Switzerland

"In rheumatology during my time, we had good people and discovered some interesting things. We were looking for how joints get destroyed in rheumatoid arthritis and bone disease, and we discovered the enzymes that are involved in breaking down the joint structure, collagenases. At the time it was a big discovery."

Tribute to Stephen M. Krane. Dayer JM, Goldring MB, Goldring SR, Kronenberg HM, Martin TJ, Russell RG (May 2015). *Journal of Bone and Mineral Research*, 30(5): 751–752.

Krane SM, In: Something in the Ether: A Bicentennial History of Massachusetts General Hospital, 1811–2011 (Memoirs Unlimited) 1st Edition. This chapter is dedicated to Stephen M. Krane, MD, 1927–2015 (see footnote).

6.1 Introduction

The aim of this chapter is to take a broad view on the evolution of research in relation to cytokine biology and matrix metalloproteinases (MMPs) in the field of rheumatology. It will focus mostly on the secreted interstitial MMP-collagenases which are MMP-1, –8, and –13. These collagenases mainly cleave fibrillar collagen types I, II, and III. In addition, they exhibit specificity for other substrates such as gelatin, casein, aggrecan, laminin, versican, perlecan, fibronectin, and tenascin. Historically, the seminal observation for the production of the human collagenase (later called MMP-1) by human synovial cells from patients with rheumatoid arthritis (RA) led to the unraveling of the link between matrix degradation and the biological function of inflammatory molecules produced by immune cells.[1–3] These inflammatory molecules were later identified as interleukin-1 (IL-1) and tumor necrosis factor (TNF). These findings illustrate the seminal link between the extracellular matrix (ECM), the enzymes that degrade it, and immunology, inflammation, and tissue destruction. Therefore, the main objective of this chapter is to describe the relationship between MMPs and cytokines, with a focus on RA, osteoarthritis (OA), and fibrosis. We have focused on RA and OA because these arthritic conditions display a plethora of both cytokines and MMPs, while fibrosis, perhaps in contrast to initial expectations, also has elevated levels of MMPs. Before examining the diseases it is interesting to go back to the history for the link between MMP and cytokines.

6.2 History of MMPs and Cytokines in RA

From the 1950s until the 1970s, concomitantly with interest in the role of immune complexes in RA, biochemistry dominated the research agenda due to the available information on structural components of tissue, mainly collagens and proteoglycans. The terminology of "connective tissue diseases" was in fashion. Pioneer studies on abnormalities of the ECM prompted biochemists to unravel structural and genetic anomalies of the matrix, leading to the concept that abnormal or denatured endogenous proteins or peptides — being recognized as foreign antigens — gave rise to an autoimmune disease. Subsequently, during the 1970s–1980s, progress in enzymology prompted scientists to investigate aberrations in enzymes and proenzymes, as well as their autoactivation. Early in the seventies, investigators began to pay closer attention to cellular aspects of the synovial tissue of RA patients (pannus), mainly due to the availability of cell cultures from normal and pathological tissue.

The seminal work at the Arthritis Unit (Massachusetts General Hospital; MGH) focusing on collagen, collagenase, and tissue destruction was inspired by the principle of limb regeneration in amphibia (tadpole tissue), where interstitial collagenase (now termed MMP-1) was first isolated by Gross and colleagues.[4] The Institute of Basic Embryology (The Robert W. Lovett Memorial Group for the study of diseases causing deformities, MGH and Harvard Medical School, Boston) working on organogenesis in amphibians, catalyzed the discovery of human collagenase in cultured fragments of RA synovium and synovial fluid from patients with RA.[5,6]

We (Dayer and colleagues) found that the main cellular source of both interstitial collagenase and prostaglandin E2 (PGE2) in synovial tissue were adherent stellate fibroblast-like cells (ASCs), also called type B fibroblastic synovial lining cells, and not monocyte/macrophages.[1–3] Of interest, the cells producing most of the collagenase were also the major source of their substrate, the collagen. The presence of collagenase in these particular cells was confirmed later by immune localization.[7]

When we cultured ASC obtained from proteolytic dispersion of the synovium of RA patients, we observed that after passages in culture, the constitutive production of collagenase decreased, coinciding with the disappearance of contaminating leucocytes. Based on the hypothesis that these contaminating leucocytes were able to induce collagenase in ASC, we found that conditioned medium of cultured stimulated lymphocyte–monocytes (which did not contain detectable amounts of collagenase and PGE2) led to the production of collagenase and PGE2 by ASC. We termed the factor responsible for this biological activity "mononuclear cell factor" (MCF) (see Fig. 6.1).[8,9]

At this early stage of the discovery, investigators exchanged their soluble factors derived from leucocytes after partial purification based on their respective functional bioassays with the aim of identifying their specificity. Our purified "MCF" was found to have similar properties to a factor called "lymphocyte-activating factor" (LAF), which had an identical molecular weight of about 15 kDa and shared chromatographic and other biochemical properties.[10] It was not until 1979 that the nomenclature of IL-1 was coined at a meeting (Ermatingen, Switzerland) and defined as that previously identified by means of different bioassays by several investigators,

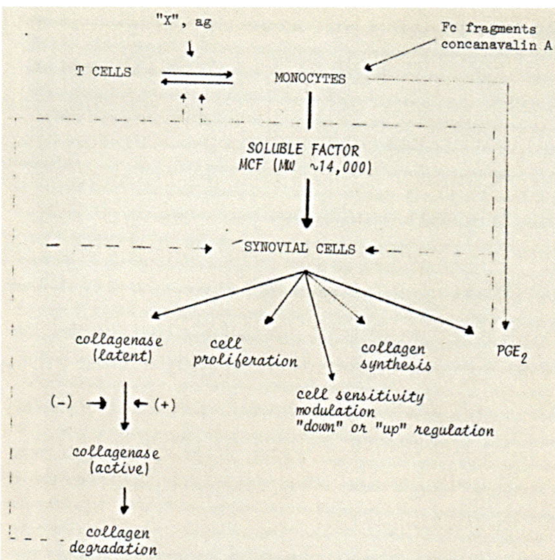

Figure 1. Possible sequences in pathophysiology of connective tissue destruction

Bull Schweiz Akad Med Wiss. 1979 Sep;35(4-6):329-43. page 340.
Collagenase and prostaglandin in connective tissue destruction: cell-cell and humoral interactions.
Dayer JM, Goldring SR, Robinson DR, Krane SM.

Figure 6.1. Possible sequences in physiopathology of connective tissue destruction.

including LAF, and the "endogenous pyrogen" (EP) that induced fever. In the field of rheumatology, other partially purified molecules, such as catabolin inducing cartilage degradation and osteoclast-activating factor (OAF) inducing calcium[45] release from bone, were also found to have similar functions to IL-1 (Table 6.1).[11] At that time, cytokines were defined only by biochemistry and not yet cloned.[12] Human IL-1 had been cloned based on its EP bioactivity, using an antibody to EP, and then recombinant IL-1 was found to have biological functions identical to those of MCF.[13] More recently, the role of IL-1 in rheumatology has been reviewed.[14]

In the course of further investigations, it came apparent that interstitial collagenase (MMP-1) and PGE2 were not induced by just one cytokine (e.g., IL-1) because we also detected biological activities in other, partially purified, culture supernatants devoid of IL-1. A molecule called cachectin (inducing cachexia), later called tumor necrosis factor-α (TNF-α), had been under close scrutiny in the fields of cancer research and infectious diseases. It was observed that cachectin/TNF-α also significantly stimulated the production of collagenase and PGE2 by fibroblasts and synovial cells from RA patients. For the first time, there was an important link between RA and TNF-α.[15] Subsequently, TNF-α was also found to stimulate bone resorption and cartilage destruction. However, IL-1 was observed to be 10 times more potent than TNF-α in inducing MMP-1 production in human synovial cells and chondrocytes. Furthermore, there is synergy between IL-1 and TNF, a finding that amplified the potential ability of two cytokines to destroy the ECM in RA.

Table 6.1. Preinterleukin-1 period.

Acronyms	Biological activities
Mononuclear cell factor (MCF)	Induction of collagenase and PGE_2 by RA synoviocytes
Osteoclast-activating factor (OAF)	Calcium[45] release from bone
Catabolin	Cartilage degradation
Endogenous pyrogen (EP)	Fever induction

6.3 Other Factors Involved in the Stimulation of MMPs

Many extracellular factors can be involved in the induction of MMPs.

They can be classified as follows:

- Endogenous soluble factors: cytokines, matrix products (matrixins), cellular microparticles, clotting factors, adhesion molecules, growth factors, collagen, and fragments[16]
- Exogenous soluble factors: bacterial, viral, fungi products mainly through toll-like receptors (TLRs), important receptors for mediating stress and danger signals
- Mechanical stress, hormones, prostanoids, adipokines
- Osmotic and ion modification
- Membrane cell-surface molecules via direct cell–cell contact[17]

There is substantial evidence that direct contact between lymphocytes and monocytes, or between lymphocytes and fibroblasts, is crucial in the induction of cytokines and MMPs. It was initially demonstrated that monocyte-macrophages and lymphocytes are crucial in the production of IL-1.[17] Direct contact between the cell membrane and surface molecules was required.[18–22] In contrast to the effect on fibroblast-synovial cells, in monocyte-macrophages the lymphocyte Th2 cell membrane factors, in association with IL-4, enhanced MMP-1 while decreasing MMP-9 production by human monocytes.[23] An imbalance between interstitial collagenase and tissue inhibitor of metalloproteinases 1 (TIMP-1) in synoviocytes and fibroblasts upon direct contact with stimulated T lymphocytes has also been observed.[24] The anti-CD3 antibody activates T cell clones to induce the production of interstitial collagenase, but not TIMP in the human cell line monocytic THP-1 cells and dermal fibroblasts.[20] In addition, PGE2, by itself, could modulate collagenase production[25] as well as hormones which modulate the rheumatoid synovial cell responses to cell–cell interactions.[26]

Other cytokines also increase production of MMPs; however, IL-1 and TNF are by far the most prominent cytokines to induce MMP-1, MMP-3, and MMP-13. However, other cytokines can also achieve this function, but mostly they reinforce the IL-1 and TNF stimulation. Among them are IL-17 and human TNF-like weak inducer of apoptosis (hTWEAK).[27,28] In contrast, IL-4 decreases MMP biosynthesis by fibroblasts.[29] A nonexhaustive list includes an important IL-6 family cytokine, the oncostatin M (OSM). OSM and IL-1 dramatically stimulate collagen release from cartilage — a key step in cartilage destruction. MMP-1, –3, –8, and –13 are all upregulated by this cytokine combination along with members of a disintegrin and metalloproteinase (ADAM) family and can be inhibited by TIMPs.[30] The proinflammatory adipokine, leptin, has a catabolic role on cartilage metabolism by upregulating proteolytic enzymes and acting synergistically with other proinflammatory stimuli. Importantly, this suggests that the infrapatellar fat pad in arthritic joints is a local producer of leptin, which may contribute to the inflammatory and degenerative processes in cartilage catabolism, and thus provide a mechanistic link between obesity and OA.[31]

As mentioned earlier, an additional cytokine that plays a role in destructive chronic arthritic diseases is hTWEAK.[28] hTWEAK synergizes with IL-1 and TNF to increase the production of PGE2, MMP-1, IL-6, and the chemokines IL-8, regulated on activation, normal T cell expressed and secreted (RANTES), and interferon-gamma-inducible protein-10 (IP-10). IL-33, an important cytokine of the IL-1 family increased the levels of proinflammatory molecules and MMPs, promoted inhibitor of kappaB (IkBα) degradation, and increased the activity of the transcription factor, nuclear factor-kB (NF-kB); these effects were reversed in fibroblast-like synoviocytes (FLS) transfected with IL-33. Stimulation with exogenous IL-33 increased RANKL and IP-10 mRNA expression.[32] These cytokines are important in bone resorption. Amphiregulin (AREG), a member of the epidermal growth factor (EGF) family, binds to the EGF receptor (EGFR), and activates downstream signaling pathways in an autocrine, paracrine, and juxtacrine manner. AREG-induced MMP-13 production requires activation of the EGFR, PI3K, Akt, and NF-kB signaling pathways.[33] Thrombin promotes MMP-13 expression through the protein kinase C delta (PKCδ) (c-Src/EGFR/PI3K/Akt/AP-1) signaling pathway in human chondrocytes.[34]

6.4 Balancing the Effect of Cytokines on MMPs Stimulation

Transforming growth factor beta (TGF-β) can have differential effects by repressing MMP-1 expression but can also induce MMP-13.[35] In OA, it is thought that Wnt signaling ultimately skews TGF-β signaling to favor ALK1 and Smad1/5/8 (with concomitant expression of MMP-13).[36] In addition, S100 proteins (a group of small Ca^{2+}-binding modulator proteins involved in cell-cycle progression, cell differentiation, and cytoskeletal–membrane interactions) are able to increase Wnt signaling. Canonical Wnts signaling in the synovium may play an important role in OA pathology. Stimulation of human OA synovium with Wnts and WISP1 increases expression of MMPs, whereas blocking Wnt signaling resulted in decreased expression of MMPs. This suggests that synovial Wnt/WISP1 might be a potential target for OA therapy.[37] Monocytes infected with *Mycobacterium tuberculosis* secreted lower levels of MMP-1 compared to MMP-9. Apparently, secretion of these MMPs is regulated by multiple upstream signaling pathways, which are independent of TIMP-1 secretion.[38] Some classical cytokines having a positive or negative effect on MMPs production are depicted in Table 6.2.

Table 6.2. Some examples of cytokines stimulating and inhibiting, focusing on MMP-1.

Stimulation	Inhibition
Alone:	
IL-1	
TNF	IL-4
In synergism	TGF-β
With IL-1 and TNF:	IFNγ
OSM	
PDGF, EGF, VEGF, bFGF	

bFGF, basic fibroblast growth factor; EGF, epidermal growth factor; IFN-γ, Interferon-γ; IL1, interleukin-1; OSM, oncostatin M; PDGF, platelet-derived growth factor; TGF-β, transforming growth factor beta; TNF, tumor necrosis factor; VEGF, vascular endothelial growth factor.

6.5 The Impact of the Discovery of IL-1 and TNF Inhibitors for the Control of MMPs

As early as 1984, before the cloning of IL-1, we suspected the presence of a potential natural inhibitor to IL-1. Using the bioassay of stimulation of collagenase and prostaglandin, we failed to detect IL-1 biological activities in the serum or urine of febrile patients. Based on this observation, in 1984 we found after biochemical purification that IL-1 was masked by a factor of approximately 17 kDa in the urine of febrile patients with monocytic leukemia and juvenile rheumatoid arthritis (JRA). The seminal observation was presented in 1984 at the Fourth International Lymphokine Workshop.[39–42] This factor specifically blocked the biological activities of IL-1, without affecting those of TNF-α. Studying the mechanism, it was possible to demonstrate for the first time in the cytokine field that a natural antagonist could block the binding of a cytokine from the same family to its receptor. This was the first description for the concept of interleukin-1 receptor antagonist (IL-1Ra).[43,44]

In 1986, still using urine from patients, we realized that other fractions of urine, which did not contain IL-1Ra, blocked the stimulation of collagenase and PGE2 when induced by TNF-α, but not by IL-1.[45] We further determined that this inhibitory molecule bound to TNF-α and subsequently that it was a fragment of the TNF receptor.[46–50]

It was obvious that, due to similarities between IL-1 and TNF, for example, MMP induction in synovial cells, blocking TNF either by antibodies or soluble receptors could constitute a promising therapeutic approach. Indeed, it is now clear that blocking TNF, and not IL-1, is more effective in many chronic inflammatory destructive diseases such as RA. However, blocking IL-1 is also a major therapeutic approach in many diseases linked to inflammasome pathway.[51] Blocking the inflammatory activities of these two potent cytokines has become the conceptual basis for the biological therapies currently in vogue for treating RA and other autoimmune inflammatory diseases. Since MMPs are one of the downstream targets that are affected by signaling pathways activated by IL-1 and

TNF, these antagonists can reduce the rampant connective tissue destruction seen in these diseases.

6.6 The Balance between Cytokine Activities in Controlling Inflammation

The balance between cytokines with pro- or anti-inflammatory activities is important in controlling the induction of MMPs. Some of them are particularly important and act by affecting the synthesis of MMPs, instead of blocking their enzymatic activities. For example, IP-10 (an IFN-gamma-inducible protein, which belongs to a gene family with chemotactic and mitogenic proteins associated with inflammation and proliferation) inhibits production of MMPs and stimulates production of TIMP-1 in human mononuclear phagocytes.[52] Interferon-β (IFN-β) (a form of interferon that is produced by fibroblasts that has antiviral effects and that is used in the treatment of multiple sclerosis) inhibits the ability of T lymphocytes to induce TNF and IL-1β production in monocytes upon direct cell–cell contact.[53,54] *In vivo* IFN-β treatment of synovial inflammation decreases expression of MMPs in patients with RA. IL-4 suppresses metalloproteinase biosynthesis in human alveolar macrophages.[29] Hormone such as retinoic acid can also inhibit the synthesis of collagenase.[55]

6.7 MMPs as Monitors of Arthritic Diseases in Tissue and Biological Fluids

Many studies have examined the relationship of MMPs to disease activity in RA. We found that a relative deficiency in TIMP-1, at the local level, would favor unopposed MMP activity, which might result in loss of periarticular bone matrix.[56] Furthermore, although proMMP-3 correlated closely at all time points with C-reactive protein (CRP), a mediator of general inflammation, it gave little or no additional clinical information regarding inflammation or radiographic progression. IL-1Ra and TIMP-1 showed weaker, acute-phase-like variation, which may reflect pathogenic agonist/inhibitor imbalance in the evolution of RA. Cartilage oligomeric matrix protein (COMP which enhances osteogenesis) in contrast did not reflect the inflammatory CRP-related component of the disease or the destructive aspect in this study.[57]

Of the MMPs, MMP-3 is of particular importance because it degrades collagen types II, III, IV, IX, and X, proteoglycans, fibronectin, laminin, and elastin. In addition, MMP-3 can also activate other MMPs such as MMP-1, MMP-7, and MMP-9, rendering MMP-3 crucial in connective tissue destruction and remodeling.

In a cohort of patients with RA and symptoms onset of <2 years, multivariate analysis identified anti-cyclic citrullinated peptide (anti-CCP; autoantibodies against citrullinated proteins in RA patients) status and baseline MMP-3 as the strongest independent predictors of radiographic disease outcome at 8.2 years. Elevated serum MMP-3 for 3–6 months predicted 1-year radiographic progression in RA, according to a small prospective cohort study.[58] These findings suggest that determination of baseline MMP-3, in conjunction with traditional serologic markers, may provide additional prognostic information for patients with RA. In other studies, enhanced production of TIMPs by peripheral blood mononuclear cells was found in RA patients responding to methotrexate treatment.[59] All of these investigations highlight the importance of continued research into a broad range of MMPs that may serve as biomarkers in predicting joint damage in RA.[60]

In healthy adults, progressive loss of cartilage matrix and cellularity occurs with age. This is accompanied with increased levels of oxidative stress, apoptosis, MMP-13, and with a decrease in chondrocyte autophagy. These changes explain the marked predisposition of joints to develop OA with age. Computational modeling provides useful insights into the underlying mechanisms involved in the age-related changes in musculoskeletal tissues. Oxidative changes and signaling pathways are pivotal in initiating age-related changes in articular cartilage.[61] Serpin peptidase inhibitor clade E member 2 (SERPINE2) might prevent cartilage catabolism by inhibiting the expression of MMP-13, one of the most relevant collagenases, involved in cartilage breakdown in OA.[62]

Using near-infrared fluorescence (NIRF) probes activated by MMPs, it is possible to visualize OA progression starting from its early stages *in vivo*. Imaging of MMP activity in an OA mouse model provided sensitive and consistent visualization of OA progression, beginning from the early stages of OA. In addition to facilitating the preclinical study of OA modulators, this approach has the potential for future human translation.[63]

6.8 Regulation of MMPs by MicroRNA in OA

A large body of literature now indicates that some specific microRNAs (miRNAs) are involved in the control of MMPs, particularly in OA. miRNA-9 increase is associated with MMP-13,[64] while MMP-13 expression inversely correlated with miRNA-27b expression.[65] There was a significant reduction in miR-127-5p expression in OA cartilage compared to normal cartilage. Upregulation of MMP-13 expression by IL-1β was correlated with downregulation of miR-127-5p expression in human chondrocytes. MicroRNA-127-5p suppressed IL-1β-induced MMP-13 production as well as the activity of a reporter construct containing the 3′-untranslated region (3′-UTR) of human MMP-13 mRNA. In addition, mutation of the miR-127-5p binding site in the 3′-UTR of MMP-13 mRNA abolished miR-127-5p-mediated repression of reporter activity. Conversely, treatment with anti-miR-127-5p remarkably increased reporter activity and MMP-13 production. Interestingly, the IL-1β-induced activation of JNK, p38, and NF-kB and expression of MMP-1 and cyclooxygenase 2 (COX2) were significantly inhibited by miR-127-5p.[66] Chondrocytes from OA patients also showed a decrease in the expression of miRNA-148a, while its overexpression inhibited the presence of Col10A1, MMP-13, and ADAMTS-5. Therefore, different approaches that increase miRNA-148a have been suggested to inhibit chondrocyte hypertrophy.[67]

6.9 Synovial Tissue MMPs in RA and OA

RA is characterized by irreversible joint destruction and results in considerable disability in severely affected individuals.[68] The joint destruction seen in RA is a consequence of an inflammatory microenvironment produced by effector cells and cytokines.[69] Traditionally, research has focused around hematopoietic cells and their subsequent production of inflammatory cytokines, such as IL-6 and TNF, for which targeted biologic therapy has shown success in slowing the progression of disease.[70] However, recent research has shown the importance of mesenchymal cells within the synovium in driving and maintaining the inflammatory microenvironment. Fibroblasts have now been shown to have an active role in the inflammatory cascade, with activated rheumatoid arthritis synovial fibroblasts (RASFs) producing MMPs and cathepsins; causing bone and cartilage destruction.[71,72] MMPs have been shown to be essential for destruction of articular matrix by cleaving ECM components. However, they have also been shown to cleave each other, resulting in a cascade of MMP activation and joint destruction.[72]

Detectable levels of almost all MMPs are found in the synovium of RA, with MMP-1 the most abundant. MMP-3 has been shown to be correlated with "invasiveness" of joint disease, with MMP-3 shown to have a role in promoting migration of inflammatory cells across the basement membrane.[72] Furthermore, evidence of advancement in radiographic RA has been shown to correlate with raised serum MMP-3 levels.[60] MMP-3 has been demonstrated as a reliable marker of disease severity in juvenile idiopathic arthritis (JIA), RA, and ankylosing spondylitis along with traditional markers such as erythrocyte sedimentation rate (ESR) and CRP.[73,74] Fibroblasts have been shown to produce MMP-3, along with other inflammatory cytokines due to activation by galectin-3, an animal lectin present at sites of destruction within the synovium.[75] As well as directly producing MMPs, fibroblasts within the synovium have also been shown to be responsible for the activation of hematopoietic cells, such as natural killer (NK) cells, resulting in increased levels of IL-6, IL-8, IL-15, and MMP-3.[76]

Histone methylation is also involved in the elevated expression of the MMPs 1, 3, 9, and 13, which contributes to cartilage destruction in RA. The alteration of histone methylation patterns, such as H3K4me3 and H3K27me3, gives rise to the change in the chromatin structure in MMP genes. Consequently, the binding of the IL-6-induced transcription factor to the promoters causes MMP gene activation. In part, this study helps to clarify some of the mechanisms regulating the activated phenotypes of RASFs at the transcriptional level.[77] There have been many reviews on this subject.[78] Epigenetic studies have shown that RASFs have altered expression of MiR-203, with hypomethylation of the promotor region causing increased production of IL-6 and MMP-1 transcripts by twofold.[79] MMPs have also been suggested as a driver of angiogenesis in the synovium, therefore easing accessibility of the joint space to migrating hematopoietic cells.[80] Finally, "inflammatory foci" within osteoarthritic synovium contain proinflammatory cytokines such as IL-17, IL-23, and MMP-9.[81] However, levels of inflammation within the synovium are much less marked in OA compared with that of RA.[78] The osteoimmunology aspect of OA has been recently reviewed.[82]

6.10 MMPs and Fibrosis

Fibrosis is the excessive deposition of fibrous connective tissue in an organ or tissue. Fibrosis can happen during a reparative or reactive process and can be benign or an indication of a pathological condition. In pathological conditions, fibrosis is characterized by the accumulation of ECM proteins, resulting in scarring and thickening of the affected tissue. Thus, it can be considered an exaggerated wound healing response, which interferes with normal organ function.

In its simplest definition, fibrosis is the result of an imbalance between deposition and degradation of the ECM in favor of deposition. As such, one would think that in fibrotic disorders MMPs and their capacity to degrade ECM components are deficient. However, the available data provide a different image that illustrates a complex relationship between the expression of various MMPs and initiation, progression, and eventual termination of fibrotic disorders. Importantly, then, whether a given MMP will be antifibrotic or profibrotic may depend on the timing of its production as well as the type of cell producing the MMPs. Ironically, but perhaps nonsurprisingly given their pleiotropic behavior, the majority of MMPs seem to favor rather than inhibit fibrosis.[83–85]

Evidence generated in human studies has demonstrated that MMP-1, MMP-2, MMP-3, MMP-7, MMP-8, MMP-9, MMP-13, and MMP-19 levels are all increased in the bronchoalveolar lavage fluid and/or lung extracts of individuals affected by idiopathic pulmonary fibrosis, when compared to controls. This evidence clearly emphasizes the importance of MMPs in fibrotic processes and indicates their potential use as biomarkers. These increased levels, however, do not indicate whether MMPs act in pro- or antifibrotic manner. The producing cells have been identified for some but not all of these MMPs and variably include macrophages, epithelial cells, and, more rarely, myofibroblasts.

The mechanisms linking these MMPs to profibrotic activities have been partially explored in models of lung, liver, kidney, and in some circumstances skin fibrosis. As an example, MMP-3 by activating latent TGF-β through the release of latent TGF-β homodimer from both latency-associated peptide and latent TGF-β binding protein-1 may favor fibrosis.[86] Similarly, MMP-7 proteolytically activates heparin-binding epidermal growth factor (pro-HB-EGF) precursor to release active HB-EGF, which promotes human lung fibroblast proliferation. Additional mechanisms may involve the capacity of MMPs to modulate the biological activity of antifibrotic mediators. For instance, MMP-8 reduces the availability of the chemokine IP-10 (CXCL10), which has direct antifibrotic properties,[87,88] simultaneously reducing the chemotactic influx of inflammatory cells, namely macrophages, which then may not sustain the development of fibrosis. Acting in a different manner, MMP-19 induces COX2 activity, thus enhancing the production of PGE2, which also has antifibrotic properties.

More complex appears to be the scenario linking MMP-12 to profibrotic activities. This may result from the capacity of MMP-12 driven by IL-13 to promote the expression of the decoy IL-13Rα2. This acts as a negative regulator of MMP-13, which has a direct antifibrotic role.[89] Thus, the upregulation of MMP-12 finally results in downregulation of the antifibrotic activity of MMP-13 in the lung. Further mechanisms possibly involved in the profibrotic roles of MMPs in lung fibrosis may include effects on epithelial cells with

MMP-3 and MMP-7, which inhibit epithelial cell repair while favoring apoptosis. However, it should be stressed that MMP-13 produced by macrophages[90] (and MMP-19) promote liver fibrosis, in contrast to their capacity to restraining lung fibrosis. This finding again underscores the variable effect of MMPs according to the main environment where they operate.[91]

Among the antifibrotic MMPs, MMP-1 may act at several levels: enhancing epithelial proliferation and resistance to apoptosis,[92] and reversing fibrosis as demonstrated when human MMP-1 is hyperexpressed in rats with liver fibrosis.[93] However, this last experimental approach has limitations, since the amount of MMP-1 largely exceeds the physiological levels potentially resulting in lack of specificity for substrates. Furthermore, another important point is that in these settings MMP-1 is produced by cells that normally do not produce it, clearly subverting its physiological role.

The enzymatic activity of MMPs is regulated by TIMP, four of which are known. In murine models of lung fibrosis, TIMP-1 and TIMP-2 single knock out do not appear to modify the fibrotic response.[84] Interestingly, TIMP-3-/- lungs, despite an overall increase in metalloproteinase activity[94] show enhanced fibrosis. However, inflammation is increased in both TIMP-1 and TIMP-3 knockouts,[94,95] thus supporting the intriguing concept that MMPs may function predominately to control first the immune responses rather than directly influencing fibroblasts for ECM synthesis. Of further complexity appears to be the role of TIMP-1 in CCl4-induced liver fibrosis with divergent evidence indicating that liver-specific overexpression of TIMP-1 but also its deletion, result in enhanced fibrosis with no direct effect on collagen synthesis.[96–98] On the whole, however, TIMPs appear to contribute to liver fibrosis.[83]

6.11 Systemic Sclerosis and MMPs

Among human diseases characterized by pathological fibrosis of the skin and internal organs, systemic sclerosis (SSc) stands out for the complex imbrication of inflammation, vasculopathy, and autoimmunity thought to cooperatively induce the deregulated deposition of ECM. Again, in this disorder gene expression and/or protein levels of several MMPs are increased. This is the case for MMP-7, MMP-9, and MMP-12. Higher serum MMP-7 levels were found in patients with SSc when compared with controls, particularly in those with lung fibrosis.[99] Similarly, total MMP-9 and pro-MMP-9 levels were significantly elevated in SSc patients with interstitial lung disease (ILD) compared to levels in SSc patients without ILD and healthy controls.[100] Furthermore, the levels of MMP-9 correlated well with the degree of skin involvement as determined by the Rodnan score and serum concentrations of TGF-β.[101] MMP-12 levels were increased in patients with SSc and associated with severity of skin and pulmonary fibrosis and with peripheral vascular damage.[102] Furthermore, TIMP-1 expression was significantly upregulated in patient's skin. Conversely, MMP-1 expression was significantly decreased in the skin of SSc patients compared to controls.[103] Similarly, MMP-13 levels in patients with SSc were significantly lower than those in normal controls. Impressively, disease duration prior to the diagnosis was significantly shorter in SSc patients with decreased serum MMP-13 levels than in those with normal levels.[104] Along the same lines of evidence, TIMP-1 levels were significantly raised in SSc compared with normal controls and TIMP-1 levels were significantly higher in early than late disease.[105] Also the serum TIMP-2 levels were elevated in some patients with SSc and were significantly higher than those of the healthy controls and their levels were significantly correlated with the extent of skin fibrosis.[106]

Peculiar to SSc, antibodies directed against MMP-1 and MMP-3 have been described. These antibodies are suggested to inhibit MMP-1 and MMP-3 and are, therefore, proposed to be involved in the development of the fibrosis.[107,108] On the other hand, autoantibodies present in the serum of a substantial number of SSc individuals react against dermal fibroblast cell membrane and in particular against TLR-4. These have been shown *in vitro* to induce the production of MMP-1 along with other inflammatory mediators. These findings suggest that some of the autoimmune features accompanying SSc may constrain rather than enhance ECM deposition.[109,110] Furthermore, contact-dependent activation of dermal fibroblasts by

T cells, whether of the Th1, Th2, or Th17 cell subset, favors the production of MMP-1 over type I collagen, thereby providing evidence for a general paradigm in which T cells, when in the proximity of fibroblasts, may favor ECM degradation.[111–114]

In summary, MMPs and TIMPs play important roles in the development and/or resolution of fibrosis. Their serum or tissue levels are increased in human pathologies associated with skin, lung, liver, or kidney fibrosis. Most of the mechanistic studies conducted in murine models of fibrosis have provided evidence indicating a promoting rather inhibitory role of MMPs in fibrosis development. The cellular and molecular details underlying these capacities are multiple, and include the enhancement of epithelial-to-mesenchymal transition; the increased production or activation of profibrotic mediators as well as the reduction of antifibrotic mediators. Some MMPs promote abnormal epithelial cell migration or aberrant repair processes; some favor the switch of macrophage phenotypes toward the profibrotic M2 type. However, in particular MMP-1 and MMP-13 have the potential to limit fibrotic responses to injury. Figure 6.2 summarizes the many and sometime contradictory findings reported in the literature on the role of MMPs in fibrosis.[83]

6.12 Therapeutic Approaches to MMPs in Rheumatology

Traditional disease modifying agents used to treat RA have been shown to decrease MMP levels, along with other inflammatory cytokines. Serial serum samples taken from RA patients treated with leflunomide have shown reduction in clinical symptoms, which correlated with a reduction in MMP-1, MMP-3, IL-6, and IL-10.[115] Similarly, patients receiving infliximab plus methotrexate have shown a larger decrease in MMP-3 levels than those treated with methotrexate alone.[116] These findings demonstrate the importance of MMP reduction in controlling disease, but the failure to directly and specifically target MMPs can result in common side effects, such as immunosuppression, as well as liver and hematological abnormalities.

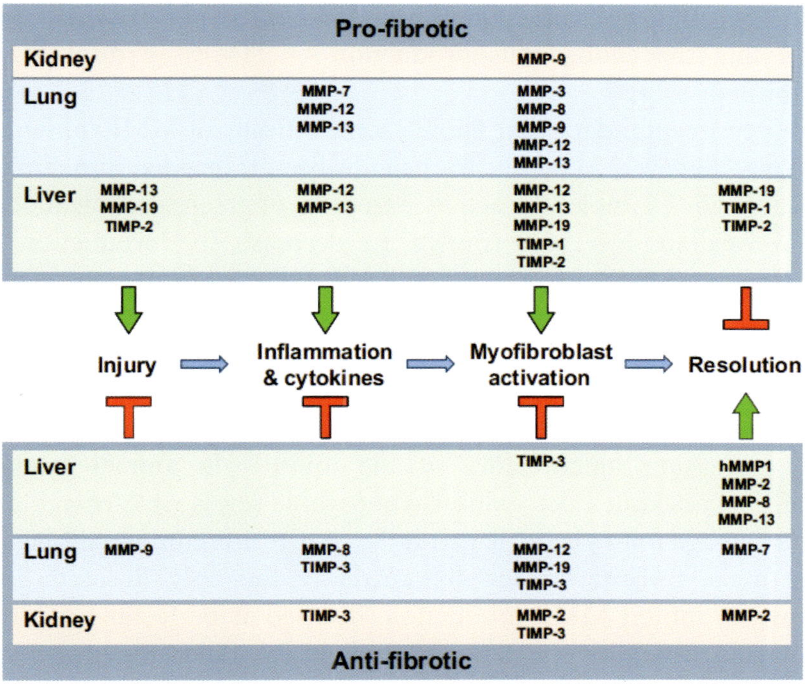

Figure 6.2. Schematic representation of the role of MMPs and TIMPs in animal models of fibrosis as reported in Giannandrea and Parks.[83] Green arrows denote enhanced activity. Red T-bars denote inhibitory activity.

Directly targeting MMPs to decrease matrix degradation and slow the development of joint destruction remains an exciting prospect in treating rheumatological diseases. Early clinical trials involving nonselective MMP inhibitors in RA were halted early due to an apparent low efficacy of the drug and many patients reported severe musculoskeletal pain and tendonitis as side effects. Lack of efficacy has been explained by poor oral bioavailability and the timing within the disease process when these drugs were used. It has been suggested that MMP inhibitors would be more efficacious in the early stages of RA, as they are key in the destruction of the joint.[117] Possible strategies for targeting MMPs include impeding the production of MMPs at the level of gene expression, blocking the active site of MMPs, and increasing the endogenous production of TIMPs (natural MMP inhibitors).

Experimental data have shown some efficacy of MMP inhibitors *in vitro*, however, as of yet this has not been translated into clinical efficacy. For example, the active ingredient of green tea, epigallocatechin-3-gallate has been shown to decrease the production of MMP-3 and MMP-1 from human RASF collected from arthroplasty patients with advanced RA. However, poor bioavailability limits this approach as a viable treatment strategy.[118] In addition, chondroitin sulfate, a "chondroprotective drug" has been shown to decrease levels of MMPs in OA, resulting in reduced levels of pain and increasing functionality of the joint.[119] In animal models, MMP-3 knockout mice exhibit comparable levels of cartilage and bone damage following collagen-induced arthritis compared to wild-type mice.[120] Tetracyclines are currently the only licensed drug known to inhibit MMP production, licensed as periostat for periodontal disease. However, clinical trials in RA failed to show a significant difference in the progression of joint damage when compared to placebo.[121]

Data from experimental systems that increase the levels of TIMPs have been more promising. Overexpression of TIMP-1, TIMP-3, and TIMP-4 inhibit inflammation in RA-like animal models.[122] Conversely, TIMP-3 knockout mice display enhanced inflammation and cartilage destruction. TIMP-3 blocks aggrecanases and TNF-α converting enzyme *in vitro* and has the potential to reduce synovial proliferation, its invasion of cartilage and inflammation *in vivo*, with this area of research ongoing.[122]

In conclusion, the early seminal work documented the pivotal role of MMPs as mediators of ECM destruction in RA. Subsequent investigations have focused on identifying the cytokines and inflammatory/immunological molecules that act in concert with MMPs in RA, OA, and fibrotic diseases. It has become increasingly clear that there is an exceedingly complex network of pathways and mechanisms that can enhance or inhibit matrix degradation. The balance between these two possible outcomes may depend on the stage of disease and treatment regimes. Although some therapies are currently available that reduce destruction of the ECM, these therapies are not specific to MMPs. Therefore, the search for targeted agents that reduce particular MMPs at particular stages of disease remains an important goal. In the meantime, however, MMPs, in particular MMP-3 (stromelysin) in RA, and TIMPs may serve as important biomarkers to help track the progression, or lack thereof, of rheumatic diseases.

References

1. Dayer, JM, SM Krane, RG Russell and DR Robinson (March 1976). Production of collagenase and prostaglandins by isolated adherent rheumatoid synovial cells. *Proceeding of the National Academy of Science USA*, 73(3), 945–949.
2. Dayer, JM, R Graham, G Russell and SM Krane (14 January 1977). Collagenase production by rheumatoid synovial cells: stimulation by a human lymphocyte factor. *Science*, 195(4274), 181–183.
3. Dayer, JM, DR Robinson and SM Krane (1 May 1977). Prostaglandin production by rheumatoid synovial cells: stimulation by a factor from human mononuclear cells. *Journal of Experimental Medicine*, 145(5), 1399–1404.
4. Gross, J and CM Lapiere (1962). Collagenolytic activity in amphibian tissues: a tissue culture assay. *Proceedings of the National Academy of Sciences of the USA*, 48, 1014–1022.
5. Evanson, JM, J Jeffrey and SM Krane (1967). Human collagenase: identification and characterization of an enzyme from rheumatoid synovium in culture. *Science*, 158, 499–502.
6. Harris, Jr. ED, R DiBona and SM Krane (1969). Collagenases in human synovial fluid. *Journal of Clinical Investigation*, 48, 2104–2113.
7. Woolley, DE, MD Crossley and JM Evanson (1977). Collagenase at sites of cartilage erosion in the rheumatoid joint. *Arthritis and Rheumatism*, 20, 1231–1239.
8. Dayer, JM, SR Goldring, DR Robinson and SM Krane (September 1979). Collagenase and prostaglandin in connective tissue destruction: cell-cell and humoral interactions. *Bull Schweiz Akad Med Wiss*, 35(4–6), 329–343.

9. Dayer, JM, ML Stephenson, E Schmidt, W Karge and SM Krane (23 February 1981). Purification of a factor from human blood monocyte-macrophages which stimulates the production of collagenase and prostaglandin E2 by cells cultured from rheumatoid synovial tissues. *FEBS Letters*, 124(2), 253.
10. Mizel, SB, JM Dayer, SM Krane and SE Mergenhagen (April 1981). Stimulation of rheumatoid synovial cell collagenase and prostaglandin production by partially purified lymphocyte-activating factor (interleukin 1). *Proceeding of National Academy Science USA*, 78(4), 2474–2477.
11. Aarden, L *et al.* (1979). Revised nomenclature for antigen nonspecific T cell proliferation and helper factors (letter) L.A. *Journal of Immunology*, 123, 2928–2929.
12. Dayer, JM (February, 2004). The process of identifying and understanding cytokines: from basic studies to treating rheumatic diseases. *Best Practice and Research Clinical Rheumatology*, 18(1), 31–45. Review.
13. Dayer, JM, B de Rochemonteix, B Burrus, S Demczuk, CA Dinarello (February 1986). Human recombinant interleukin 1 stimulates collagenase and prostaglandin E2 production by human synovial cells. *Journal of Clinical Investigation*, 77(2), 645–648.
14. Schett, G, JM Dayer and B Manger (January, 2016). Interleukin-1 function and role in rheumatic disease. *Nature Reviews Rheumatology*, 12(1), 14–24.
15. Dayer, JM, B Beutler and A Cerami (1 December 1985). Cachectin/tumor necrosis factor stimulates collagenase and prostaglandin E2 production by human synovial cells and dermal fibroblasts. *Journal of Experimental Medicine*, 162(6), 2163–2168.
16. Biswas, C and JM Dayer (December 1979). Stimulation of collagenase production by collagen in mammalian cell cultures. *Cell*, 18(4), 1035–1041.
17. Dayer, JM, J Bréard, L Chess and SM Krane (November 1979) Participation of monocyte-macrophages and lymphocytes in the production of a factor that stimulates collagenase and prostaglandin release by rheumatoid synovial cells. *Journal of Clinical Investigation*, 64(5), 1386–1392.
18. Vey, E, JH Zhang and JM Dayer (15 September 1992). IFN-gamma and 1, 25(OH) 2D3 induce on THP-1 cells distinct patterns of cell

surface antigen expression, cytokine production, and responsiveness to contact with activated T cells. *Journal Immunology*, 149(6), 2040–2046.
19. Lacraz, S, P Isler, E Vey, HG Welgus and JM Dayer (2 September 1994). Direct contact between T lymphocytes and monocytes is a major pathway for induction of metalloproteinase expression. *Journal of Biological Chemistry*, 269(35), 22027–22033.
20. Miltenburg, AM, S Lacraz, HG Welgus and JM Dayer (15 March 1995). Immobilized anti-CD3 antibody activates T cell clones to induce the production of interstitial collagenase, but not tissue inhibitor of metalloproteinases, in monocytic THP-1 cells and dermal fibroblasts. *Journal of Immunology*, 154(6), 2655–2667.
21. Dayer, JM and D Burger (June 2004). Cytokines and direct cell contact in synovitis: relevance to therapeutic intervention. *Autoimmunity Reviews,* 3(Suppl. 1), S14.
22. Dayer, JM and D Burger (1999). Cytokines and direct cell contact in synovitis: relevance to therapeutic intervention. *Arthritis Research*, 1(1), 17–20.
23. Chizzolini, C, R Rezzonico, C, De Luca, D Burger and JM Dayer (1 June 2000). Th2 cell membrane factors in association with IL-4 enhance matrix metalloproteinase-1 (MMP-1) while decreasing MMP-9 production by granulocyte-macrophage colony-stimulating factor-differentiated human monocytes. *Journal of Immunology*, 164(11), 5952–5960.
24. Burger, D, R Rezzonico, JM Li, C Modoux, RA Pierce, HG Welgus and JM Dayer (October 1998). Imbalance between interstitial collagenase and tissue inhibitor of metalloproteinases 1 in synoviocytes and fibroblasts upon direct contact with stimulated T lymphocytes: involvement of membrane-associated cytokines. *Arthritis and Rheumatology*, 41(10), 1748–1759.
25. Dayer, JM, MS Roelke and SM Krane (15 September 1984). Effects of prostaglandin E2, indomethacin, trifluoperazine and drugs affecting the cytoskeleton on collagenase production by cultured adherent rheumatoid synovial cells. *Biochemical Pharmacology*, 1984, 33(18), 2893–2899.
26. Goldring, SR, JM Dayer and SM Krane (March 1984). Rheumatoid synovial cell hormone responses modulated by cell-cell interactions. *Inflammation*, 8(1), 107–121.

27. Chabaud, M, P Garnero, JM Dayer, PA Guerne, F Fossiez and P Miossec (July 2000). Contribution of interleukin 17 to synovium matrix destruction in rheumatoid arthritis. *Cytokine*, 12(7), 1092–1099.
28. Chicheportiche, Y, R Chicheportiche, I Sizing, J Thompson, CB Benjamin, C Ambrose and JM Dayer (2002). Proinflammatory activity of TWEAK on human dermal fibroblasts and synoviocytes: Blocking and enhancing effects of anti-TWEAK monoclonal antibodies. *Arthritis Research*, 4(2), 126–133.
29. Lacraz, S, L Nicod, B Galve-de Rochemonteix, C Baumberger, JM Dayer and HG Welgus (August 1992). Suppression of metalloproteinase biosynthesis in human alveolar macrophages by interleukin-4. *Journal of Clinical Investigation*, 90(2), 382–388.
30. Cawston, TE and DA Young (January 2010). Proteinases involved in matrix turnover during cartilage and bone breakdown. *Cell and Tissue Research*, 339(1), 221–235.
31. Hui, W, GJ Litherland, MS Elias, GI Kitson, TE Cawston, AD Rowan and DA Young (March 2012). Leptin produced by joint white adipose tissue induces cartilage degradation via upregulation and activation of matrix metalloproteinases. *Annals of the Rheumatic Diseases*, 71(3), 455–462.
32. Lee, EJ, MW So, S Hong, YG Kim, B Yoo and CK Lee (January 2016). Interleukin-33 acts as a transcriptional repressor and extracellular cytokine in fibroblast-like synoviocytes in patients with rheumatoid arthritis. *Cytokine*, 77, 35–43.
33. Chen, YT, CH Hou, SM Hou and JF Liu (2014). The effects of amphiregulin induced MMP-13 production in human osteoarthritis synovial fibroblast. *Mediators of Inflammation*, 2014, 759028.
34. Huang, CY, HJ Lin, HS Chen, SY Cheng, HC Hsu and CH Tang (2013). Thrombin promotes matrix metalloproteinase-13 expression through the PKCδ c-Src/EGFR/PI3K/Akt/AP-1 signaling pathway in human chondrocytes. *Mediators of Inflammation*, 2013, 326041.
35. Uría, JA, MG Jiménez, M Balbín, JMP Freije and C López-Otín (1998). Differential effects of transforming growth factor-β on the expression of collagenase-1 and collagenase-3 in human fibroblasts. *Journal of Biological Chemistry*, 273(16), 9769–9777.
36. van den Boschm, MH, AB Blom, PL van Lent, HM van Beuningen, EN Blaney Davidson, M van der Kraan and WB van den Berg

(May 2014). Canonical Wnt signaling skews TGF-β signaling in chondrocytes towards signaling via ALK1 and Smad 1/5/8. *Cellular Signalling*, 26(5), 951–958.
37. van den Bosch, MH, AB Blom, RF Schelbergen, T Vogl, JP Roth, AW Slöetjes, WB van den Berg, PM van der Kraan and PL van Lent (January 2016). Induction of canonical Wnt signaling by the alarmins S100A8/A9 in murine knee joints: implications for osteoarthritis. *Arthritis and Rheumatology*, 68(1), 152–163.
38. Friedland, JS, TC Shaw, NM Price and JM Dayer (January 2002). Differential regulation of MMP-1/9 and TIMP-1 secretion in human monocytic cells in response to *Mycobacterium tuberculosis*. *Matrix Biology: Journal of the International Society for Matrix Biology*, 21(1), 103–110.
39. Balavoine, JF, B de Rochemonteix, A Cruchaud and JM Dayer (1984). Identification of interleukin 1-like activity and inhibitor(s) in urine from a patient with acute monoblastic leukemia. In *4th International Lymphokine Workshop*. October 17–21, Germany: Schloss Elmau. *Lymphokine Research*, 3, 23 (abstract).
40. Dayer, JM (1985). Collagenase- and PGE2-stimulating activity (Interleukin-1-Like) and inhibitor in urine from a patient with monocytic leukemia. In *The Physiologic, Metabolic, and Immunologic Actions of Interleukin-1, Progress in Leukocyte Biology*, JK Matthew, JJ Oppenheim and MC Powanda, (eds.), vol. 2, p. 429. New York, NY: Alan R. Liss.
41. Balavoine, JF, B de Rochemonteix, K Williamson, P Seckinger, A Cruchaud and JM Dayer (October 1986). Prostaglandin E2 and collagenase production by fibroblasts and synovial cells is regulated by urine-derived human interleukin 1 and inhibitor(s). *Journal of Clinical Investigation*, 78(4), 1120–1124.
42. Prieur, AM, MT Kaufmann, C Griscelli and JM Dayer (28 November 1987). Specific interleukin-1 inhibitor in serum and urine of children with systemic juvenile chronic arthritis. *Lancet*, 2(8570), 1240–1242.
43. Seckinger, P, JM Lowenthal, K Williamson, JM Dayer and HR MacDonald (1 September 1987). A urine inhibitor of interleukin 1 activity that blocks ligand binding. *Journal of Immunology*, 139(5), 1546–1549.
44. Seckinger, P, K Williamson, JF Balavoine, B Mach, G Mazzei, A Shaw and JM Dayer (1 September 1987). A urine inhibitor of

interleukin 1 activity affects both interleukin 1 alpha and 1 beta but not tumor necrosis factor alpha. *Journal of Immunology*, 139(5), 1541–1545.

45. Seckinger, P, S Isaaz and JM Dayer (1 April 1988). A human inhibitor of tumor necrosis factor alpha. *Journal of Experimental Medicine*, 167(4), 1511–1516.
46. Seckinger, P, S Isaaz and JM Dayer (15 July 1989). Purification and biologic characterization of a specific tumor necrosis factor alpha inhibitor. *Journal of Biological Chemistry*, 264(20), 11966–11973; Mazzei, GJ, PL Seckinger, JM Dayer and AR Shaw (March 1990). Purification and characterization of a 26-kDa competitive inhibitor of interleukin. *European Journal of Immunology*, 20(3), 683–689.
47. Seckinger, P, E Vey, G Turcatti, P Wingfield and JM Dayer (May 1990). Tumor necrosis factor inhibitor: purification, NH2-terminal amino acid sequence and evidence for anti-inflammatory and immunomodulatory activities. *European Journal of Immunology*, 20(5), 1167–1174.
48. Seckinger, P, MT Kaufmann and JM Dayer (June 1990). An interleukin 1 inhibitor affects both cell-associated interleukin 1-induced T cell proliferation and PGE2/collagenase production by human dermal fibroblasts and synovial cells. *Immunobiology*, 180(4–5), 316–327.
49. Seckinger, P, JH Zhang, B Hauptmann and JM Dayer (July 1990). Characterization of a tumor necrosis factor alpha (TNF-alpha) inhibitor: evidence of immunological cross-reactivity with the TNF receptor. *Proceeding of the National Academy of Science USA*, 87(13), 5188–5192.
50. Seckinger, P, J Klein-Nulend, C Alander, RC Thompson, JM Dayer and LG Raisz. (15 December 1990). Natural and recombinant human IL-1 receptor antagonists block the effects of IL-1 on bone resorption and prostaglandin production. *Journal of Immunology*, 145(12), 4181–4184.
51. Schett, G, D Elewaut, IB McInnes, JM Dayer and MF Neurath (July 2013). How cytokine networks fuel inflammation: toward a cytokine-based disease taxonomy. *Nature Medicine*, 19(7), 822–824.
52. Lacraz, S, LP Nicod, R Chicheportiche, HG Welgus and JM Dayer (November 1995). IL-10 inhibits metalloproteinase and stimulates TIMP-1 production in human mononuclear phagocytes. *Journal of Clinical Investigation*, 96(5), 2304–2310.

53. Jungo, F, JM Dayer, C Modoux, N Hyka and D Burger (7 June 2001). IFN-beta inhibits the ability of T lymphocytes to induce TNF-alpha and IL-1beta production in monocytes upon direct cell-cell contact. *Cytokine*, 14(5), 272–282.
54. Smeets, TJ, JM Dayer, MC Kraan, J Versendaal, R Chicheportiche, FC Breedveld and PP Tak (February 2000). The effects of interferon-beta treatment of synovial inflammation and expression of metalloproteinases in patients with rheumatoid arthritis. *Arthritis and Rheumatology*, 43(2), 270–274.
55. Brinckerhoff, CE, RM McMillan, JM Dayer, ED Jr Harris (21 August 1980). Inhibition by of collagenase production in rheumatoid synovial cells. *New England Journal of Medicine*, 303(8), 432–436.
56. Murphy, E, P Roux-Lombard, T Rooney, O Fitzgerald, JM Dayer and B Bresnihan (March 2009). Serum levels of tissue inhibitor of metalloproteinase-1 and periarticular bone loss in early rheumatoid arthritis. *Clinical Rheumatology*, 28(3), 285–291.
57. Roux-Lombard, P, K Eberhardt, T Saxne, JM Dayer and FA Wollheim (May 2001). Cytokines, metalloproteinases, their inhibitors and cartilage oligomeric matrix protein: Relationship to radiological progression and inflammation in early rheumatoid arthritis. A prospective 5-year study. *Rheumatology (Oxford)*, 40(5), 544–551.
58. Ma, JD, XN Wei, DH Zheng, YQ Mo, LF Chen, X Zhang, JH Li and L Dai (2015). Continuously elevated serum matrix metalloproteinase-3 for 3–6 months predict one-year radiographic progression in rheumatoid arthritis: A prospective cohort study. *Arthritis Research and Therapy*, 17, 289.
59. Seitz, M and JM Dayer (June 2000). Enhanced production of tissue inhibitor of metalloproteinases by peripheral blood mononuclear cells of rheumatoid arthritis patients responding to methotrexate treatment. *Rheumatology (Oxford)*, 39(6), 637–645.
60. Houseman, M, C Potter, N Marshall, R Lakey, T Cawston, I Griffiths, S Young-Min and JD Isaacs (7 February 2012). Baseline serum MMP-3 levels in patients with Rheumatoid Arthritis are still independently predictive of radiographic progression in a longitudinal observational cohort at 8 years follow up. *Arthritis Research and Therapy*, 14(1):R30.
61. Hui, W, DA Young, AD Rowan, X Xu, TE Cawston and CJ Proctor (2014). Oxidative changes and signalling pathways are pivotal in

initiating age-related changes in articular cartilage. *Annals Rheumatic Diseases*, 75, 1–10.

62. Santoro, A, J Conde, M Scotece, V Abella, A Lois, V Lopez *et al.* (25 August 2015). SERPINE2 inhibits IL-1α-induced MMP-13 expression in human chondrocytes: Involvement of ERK/NF-kB/AP-1 pathways. *PLoS ONE*, 10(8). doi:10.1371/journal.pone.0135979.
63. Leahy, AA, SA Esfahani, AT Foote *et al.* (2015). Following the trajectory of osteoarthritis development through serial near infrared fluorescence imaging of MMP activities. *Arthritis and rheumatology*, 67(2), 442–453.
64. Jones, SW, G Watkins, N Le Good, S Roberts, CL Murphy, SMV Brockbank *et al.* (2009). The identification of differentially expressed microRNA in osteoarthritic tissue that modulate the production of TNF-α and MMP13. *Osteoarthritis Cartilage*, 17, 464–472.
65. Akhtar, N, Z Rasheed, S Ramamurthy, AN Anbazhagan, FR Voss, and TM Haqqi (2010). MicroRNA-27b regulates the expression of matrix metalloproteinase 13 in human osteoarthritis chondrocytes. *Arthritis and Rheumatism*, 62, 1361–1371.
66. Park, SJ, EJ Cheon, MH Lee and HA Kim (December 2013). MicroRNA-127-5p regulates matrix metalloproteinase 13 expression and interleukin-1β-induced catabolic effects in human chondrocytes. *Arthritis and Rheumatism*, 65(12), 3141–3152.
67. Vonk, LA, AHM Kragten, WJA Dhert, DBF Saris and LB Creemers (2014). Overexpression of hsa-miR-148a promotes cartilage production and inhibits cartilage degradation by osteoarthritic chondrocytes. *Osteoarthritis Cartilage*, 22, 145–153.
68. Scott, DL, K Pugner, K Kaarela, DV Doyle, A Woolf, J Holmes *et al.* (2000). The links between joint damage and disability in rheumatoid arthritis. *Rheumatology (Oxford)*, 39, 122–132.
69. Buckley, CD, D Pilling, JM Lord, AN Akbar, D Scheel-Toellner and M Salmon (2001). Fibroblasts regulate the switch from acute resolving to chronic persistent inflammation. *Trends in Immunology*, 22, 199–204.
70. McInnes, IB and G Schett (2007). Cytokines in the pathogenesis of rheumatoid arthritis. *Nature Review Immunology*, 7, 429–442.
71. Muller-Ladner, U and S Gay (2002). MMPs and rheumatoid synovial fibroblasts: Siamese twins in joint destruction? *Annals of the Rheumatic Diseases*, 61, 957–959.

72. Pap, T, U Muller-Ladner, RE Gay and S Gay (2000). Fibroblast biology. Role of synovial fibroblasts in the pathogenesis of rheumatoid arthritis. *Arthritis Res*earch, 2, 361–367.
73. Green, MJ, AKS Gough, J Devin, J Smith, P Astin, D Taylor and P Emery (2003). Serum MMP-3 and MMP-1 and progression of joint damage in early rheumatoid arthritis. *Rheumatology*, 42, 8–88.
74. Viswanath, V, A Myles, R Dayal and A Aggarwal (2011). Levels of serum matrix metalloproteinase-3 correlate with disease activity in the enthesitis-related arthritis category of juvenile idiopathic arthritis. *Journal of Rheumatology*, 38, 2482–2487.
75. Filer, A, M Bik, GA Rabinovich and CD Buckley (2009). Galectin 3 induces a distinctive pattern of cytokine and chemokine production in rheumatoid synovial fibroblasts via selective signalling pathways. *Arthritis and Rheumatism*, 60(6), 1604–1614.
76. Chan, A, A Filer, G Parsonage, S Kollnberger, R Gundle, CD Buckley and P Bowness (2008). Mediation of the proinflammatory cytokine response in rheumatoid arthritis and spondylarthritis by interactions between fibroblast-like synoviovytes and natural killer cells. *Arthritis and Rheumatism*, 58(3), 707–717.
77. Araki, Y, T Tsuzuki Wada, Y Aizaki, K Sato, K Yokota, K Fujimoto, YT Kim, H Oda, R Kurokawa and T Mimura (2015). Histone methylation and STAT3 differentially regulate IL-6-induced MMP gene activation in rheumatoid arthritis synovial fibroblasts. Arthritis and *Rheumatology*, 68, 1111–1123. doi:10.1002/art.39563
78. Ammari, M, C Jorgensen and F Apparailly (March 2013). Impact of microRNAs on the understanding and treatment of rheumatoid arthritis. *Current Opinion Rheumatology*, 25(2), 225–233.
79. Stanczyk, J, CD Buckley, PP Tak, S Gay and D Kyburz (2011). Altered expression of microRNA-203 in rheumatoid arthritis synovial fibroblasts and its role in fibroblast activation. *Arthritis and Rheumatism*, 63(2), 373–381.
80. Walsh, DA (1999). Angiogenesis and arthritis. *Rheumatology*, 38, 103–112.
81. Deligne, C, S Casulli, A Pigenet, C Bougault, L Campillo-Gimenez, G Nourissat, F Berenbaum, C Elbim, and X Houard (2015). Differential expression of interleukin-17 and interleukin-22 in inflamed and non-inflamed synovium from osteoarthritis patients. *Osteoarthritis and Cartilage*, 23, 1843–1852.

82. Goldring, MB, J-M Dayer, SR Goldring (2015). Osteoarthritis and the immune system. In: *Interactions of the Immune and Skeletal Systems Osteoimmunology*, 2nd Ed., J Lorenzo, M Horowitz, Y Choi, H Takayanagi and G Schett (eds.), Chap. 14, pp. 257–281. ISBN9780128005712. Release Date: 05 Oct 2015.
83. Giannandrea, M and WC Parks (2014). Diverse functions of matrix metalloproteinases during fibrosis. *Diseases Model and Mechanisms*, 7, 193–203.
84. Craig, VJ *et al.* (2015). Matrix metalloproteinases as therapeutic targets for idiopathic pulmonary fibrosis. *American Journal of Respiratory Cell Molecular Biology*, 53, 585–600.
85. Rohani, MG and WC Parks (2015). Matrix remodeling by MMPs during wound repair. *Matrix Bio*logy, 44–46, 113–121.
86. Maeda, S *et al.* (2002). The first stage of transforming growth factor beta1 activation is release of the large latent complex from the extracellular matrix of growth plate chondrocytes by matrix vesicle stromelysin-1 (MMP-3). *Calcified Tissue International*, 70, 54–65.
87. Jiang, D *et al.* (2004). Regulation of pulmonary fibrosis by chemokine receptor CXCR3. *Journal of Clinical Investigation*, 114, 291–299.
88. Tager, AM *et al.* (2004) Inhibition of pulmonary fibrosis by the chemokine IP-10/CXCL10. *American Journal of Respiratory Cell Molecular Biology*, 31, 395–404.
89. Madala, SK *et al.* (2010). Matrix metalloproteinase 12-deficiency augments extracellular matrix degrading metalloproteinases and attenuates IL-13-dependent fibrosis. *Journal of Immunology*, 184, 3955–3963.
90. Fallowfield, JA *et al.* (2007). Scar-associated macrophages are a major source of hepatic matrix metalloproteinase-13 and facilitate the resolution of murine hepatic fibrosis. *Journal of Immunology*, 178, 5288–5295.
91. Jirouskova, M *et al.* (2012). Hepatoprotective effect of MMP-19 deficiency in a mouse model of chronic liver fibrosis. *PLoS ONE*, 7, e46271.
92. Herrera, I *et al.* (2013). Matrix metalloproteinase (MMP)-1 induces lung alveolar epithelial cell migration and proliferation, protects from apoptosis, and represses mitochondrial oxygen consumption. *Journal of Biological Chemistry*, 288, 25964–25975.

93. Yang, C, G Hu and D Tan (1999). Effects of MMP-1 expressing plasmid on rat liver fibrosis. *Zhonghua Gan Zang Bing Za Zhi*, 7, 230–232.
94. Gill, SE *et al.* (2010). Tissue inhibitor of metalloproteinases 3 regulates resolution of inflammation following acute lung injury. *American Journal of Pathology*, 176, 64–73.
95. Kim, KH *et al.* (2005). Tissue inhibitor of metalloproteinase-1 deficiency amplifies acute lung injury in bleomycin-exposed mice. *American Journal of Respiratory Cell and Molecular Biology*, 33, 271–279.
96. Yoshiji, H *et al.* (2000). Tissue inhibitor of metalloproteinases-1 promotes liver fibrosis development in a transgenic mouse model. *Hepatology*, 32, 1248–1254.
97. Yoshiji, H *et al.* (2002) Tissue inhibitor of metalloproteinases-1 attenuates spontaneous liver fibrosis resolution in the transgenic mouse. *Hepatology*, 36, 850–860.
98. Wang, H *et al.* (2011). Tissue inhibitor of metalloproteinase 1 (TIMP-1) deficiency exacerbates carbon tetrachloride-induced liver injury and fibrosis in mice: Involvement of hepatocyte STAT3 in TIMP-1 production. *Cell and Bioscience*, 1, 14.
99. Moinzadeh, P *et al.* (2011). Elevated MMP-7 levels in patients with systemic sclerosis: Correlation with pulmonary involvement. *Experimental Dermatology*, 20, 770–773.
100. Andersen, GN *et al.* (2007). Bronchoalveolar matrix metalloproteinase 9 relates to restrictive lung function impairment in systemic sclerosis. *Respiratory Medicine*, 101, 2199–2206.
101. Kim, WU *et al.* (2005). Elevated matrix metalloproteinase-9 in patients with systemic sclerosis. *Arthritis Research and Therapy*, 7, R71–R79.
102. Manetti, M *et al.* (2012) Increased serum levels and tissue expression of matrix metalloproteinase-12 in patients with systemic sclerosis: correlation with severity of skin and pulmonary fibrosis and vascular damage. *Annals Rheumatic Diseases*, 71, 1064–1072.
103. Frost, J *et al.* (2012). Differential gene expression of MMP-1, TIMP-1 and HGF in clinically involved and uninvolved skin in South Africans with SSc. *Rheumatology (Oxford)*, 51, 1049–1052.
104. Asano, Y *et al.* (2006). Clinical significance of serum levels of matrix metalloproteinase-13 in patients with systemic sclerosis. *Rheumatology (Oxford)*, 45, 303–307.

105. Young-Min, SA *et al.* (2001). Serum TIMP-1, TIMP-2, and MMP-1 in patients with systemic sclerosis, primary Raynaud's phenomenon, and in normal controls. *Annals Rheumatic Diseases*, 60, 846–851.
106. Yazawa, N *et al.* (2000). Serum levels of tissue inhibitor of metalloproteinases 2 in patients with systemic sclerosis. *Journal of the American Academy Dermatology*, 42, 70–75.
107. Nishijima, C *et al.* (2004). Autoantibody against matrix metalloproteinase-3 in patients with systemic sclerosis. *Clinical and Experimental Immunology*, 138, 357–363.
108. Sato, S *et al.* (2003). Function blocking autoantibodies against matrix metalloproteinase-1 in patients with systemic sclerosis. *Journal of Investigation Dermatology*, 120, 542–547.
109. Chizzolini, C *et al.* (2002). Autoantibodies to fibroblasts induce a proadhesive and proinflammatory fibroblast phenotype in patients with systemic sclerosis. *Arthritis and Rheumatology*, 46, 1602–1613.
110. Fineschi, S *et al.* (2008). Antifibroblast antibodies in systemic sclerosis induce fibroblasts to produce profibrotic chemokines, with partial exploitation of toll-like receptor 4. *Arthritis and Rheumatology*, 58, 3913–3923.
111. Chizzolini, C *et al.* (1998). Inhibition of type I collagen production by dermal fibroblasts upon contact with activated T cells: Different sensitivity to inhibition between systemic sclerosis and control fibroblasts. *Arthritis and Rheumatology*, 41, 2039–2047.
112. Chizzolini, C *et al.* (2000). Th2 cell membrane factors in association with IL-4 enhance matrix metalloproteinase-1 (MMP-1) while decreasing MMP-9 production by granulocyte-macrophage colony-stimulating factor-differentiated human monocytes. *Journal of Immunology*, 164, 5952–5960.
113. Chizzolini, C *et al.* (2003). Systemic sclerosis Th2 cells inhibit collagen production by dermal fibroblasts via membrane-associated tumor necrosis factor alpha. *Arthritis and Rheumatology*, 48, 2593–2604.
114. Truchetet, ME *et al.* (2013). Interleukin-17A+ cell counts are increased in systemic sclerosis skin and their number is inversely correlated with the extent of skin involvement. *Arthritis and Rheumatology*, 65, 1347–1356.
115. Litinsky, I, D Paran, D Levartovsky, I Wigler, I Kaufman, I Yaron, M Yaron, D Caspi and O Elkayam (2006). The effects of leflunomide

on clinical parameters and serum levels of IL-6, IL-10, MMP-1 and MMP-3 in patients with resistant rheumatoid arthritis. *Cytokine*, 33, 106–110.

116 Visvanathan, S *et al.* (2007). Changes in biomarkers of inflammation and bone turnover and associations with clinical efficacy following infliximab plus methotrexate therapy in patients with early rheumatoid arthritis. *Journal of Rheumatology*, 34, 1465–1474.

117. Jackson, C, M Nguyen, J Arkell and P Sambrook (2001). Selective matrix metalloproteinase (MMP) inhibition in rheumatoid arthritis — Targeting gelatinase a activation. *Inflammation research*, 50, 183–186.

118. Lee, S *et al.* (2008). Epigallocatechin-3-gallate suppresses TNF-a-induced production of MMP-1 and -3 in rheumatoid arthritis synovial fibroblasts. *Rheumatology International*, 29, 23–29.

119. Zhang, Y, T Pizzute and M Pei (2014). Anti-inflammatory strategies in cartilage repair. *Tissue Engineering*, 20(6), 655–668.

120. Mudgett, JS, NI Hutchinson, NA Chartrain, AJ Forsyth, J McDonnell, II Singer *et al.* (1998) Susceptibility of stomelysin 1-deficient mice to collagen-induced arthritis and cartilage destruction. *Arthritis and Rheumatology*, 41, 110–121.

121. Bluhm, GB, JT Sharp, BC Tilley, GS Alarcon, SM Cooper, SR Pillemer *et al.* (1997). Radiographic results from the minocycline in rheumatoid arthritis (MIRA) trial. *Journal of Rheumatology*, 24, 1295–1302.

122. Yaqoob, H *et al.* (2008). Inhibition of matrix metalloproteinases as a feasible therapeutic target in rheumatoid arthritis. *Rheumatology Reviews*, 4(4), 288–297.

7. Matrix Metalloproteinases (MMPs) as Cancer Therapeutic Targets

Jillian Cathcart, Vincent M. Alford and Jian Cao

Division of Cancer Prevention, Department of Medicine, Stony Brook University, Stony Brook, NY, USA

7.1 Introduction

Matrix metalloproteinases (MMPs) are a family of degradative enzymes responsible for cleaving components of the basement membrane and extracellular matrix (ECM). MMPs have been implicated in a number of key normal and pathological processes, the latter including tumorigenesis, cancer cell invasion, dysregulated angiogenesis, and metastasis. While the biology of how MMPs function in cancer has been extensively researched over the past several decades, translation of the basic science into clinical applications lags behind. However, this is not to say that research is not active in this field. Understanding how MMPs directly contribute to tumor initiation and progression allows for the information to be leveraged to identify groups with higher cancer risk and thus opportunities for early detection and improved therapeutic approaches to be developed.

7.2 When MMP Levels Increase, Cancer Risk Increases

The three described phases of tumorigenesis are initiation, including genomic alterations, abnormal proliferation of the altered cells, and progressive growth and spreading of the cancer.[1] The roles for MMPs in contributing to each of these phases are well described. But the idea that overexpression of MMPs can help to lead to tumor formation is less well accepted and poorly understood. In mouse models, overexpression of either MMP-1 or MMP-7 in a tissue-specific manner leads to hyperproliferation and increased susceptibility to cancer.[2,3] Taking another approach, overexpression of either MMP-3 or MMP-14 in mammary tissue leads to spontaneous formation of breast cancer.[3–5] In a chemically induced skin cancer model, transgenic mice which lack MMP-9 exhibit decreased hyperproliferation of keratinocytes at all neoplastic stages, with fewer high-grade tumors and decreased incidence of tumor invasion.[6] Additionally, several studies have indicated that the ECM is often physically altered prior to the development of cancer. Two separate studies have clearly implicated a role for normal physiological MMP-3-mediated remodeling of the mammary ECM as priming breast tissue for breast cancer. Such remodeling occurs during development, pregnancy, and menopause, and leads to epithelial branching and endothelial morphogenesis. This results in a disruption of stromal-epithelial homeostasis, which is hypothesized to become the basis for hyperplasias and neoplasias.[4,7] Similar results in a study of Danish postmenopausal women have been found which correlated serum levels of MMP-mediated degraded type I collagen with mortality risk. This study established degraded type I collagen as being a prognostic marker for mortality, due to cancer as well as various other pathologies with this underlying circumstance, up to nine years prior to patient death.[8]

Cancer is a multifaceted disease in which a single thing cannot be pinpointed as the root cause for its development. However, nearly all cancers overexpress MMPs and take advantage of their integral functions in promoting cancer progression and aggression. Aside from remodeling the ECM, MMPs have important tumor-promoting roles such as cleavage of E-cadherin expressed on epithelial cells.

This cleavage is known to induce cell proliferation as well as the phenotypic changes associated with epithelial-to-mesenchymal transition (EMT).[9] This combined with the observation that the stromal environment appears to be frequently altered prior to cancer development[7,10] should allow us to appreciate the idea that overexpression of these matrix remodeling- and bioactive-MMPs may contribute directly to the development of cancer.

A well-established contributor to cancer formation is inflammation. Many cancers are known to arise from sites of chronic infection or inflammation. Inflammatory cells and stromal cells responding to inflammatory signals all have physiological roles during inflammation in which they produce MMPs. In a positive feedback mechanism, MMPs can activate signaling molecules which recruit additional inflammatory cells to initial site of inflammation. In the locally inflamed environment, MMPs can cleave other bioactive signals which stimulate angiogenesis, proliferation, survival, and migration of the cells in the surrounding tissue.[11] When macrophage production of MMP-9 is genetically eliminated or chemically blocked, cancer development in human papillomavirus-16 (HPV) mice is significantly reduced, despite infiltration of the diseased tissues by immune cells being unperturbed.[12] Upregulation and subsequent activation of the latent MMPs during cancer initiation can also be an indirect consequence of inflammation. DNA-binding proteins are activated in response to an inflammatory stimulus and can elicit a response in stromal cells. Of note among these transcription factors is nuclear factor-kB (NF-kB), which transactivates inflammatory, anti-apoptotic, proliferative, proangiogenic, and ECM remodeling genes. To date, NF-kB has been implicated in increasing expression of MMPs −1, −2, −3, −7, −9, −13, and −14.[13–21] In the early stages, expression of these MMPs contributes to cancer initiation in a variety of ways. However, it must be noted that MMPs have normal, protective roles and discriminating between their physiological functions and those contributing to cancer progression remains a challenge in this field.[22]

There is a wide variety of signaling molecules which, upon cleavage by MMP, stimulate protumorigenic effects. For example, several

MMPs can cleave the ECM component proteoglycan, releasing fibroblastic growth factor (FGF) and transforming growth factor-beta (TGF-β).[23–25] Cleavage of the proform of vascular endothelial growth factor (VEGF) is required for its activation. This signaling molecule has been shown to be cleaved by MMPs –3, –7, –9, and –19 to induce angiogenesis.[26] Angiogenesis has many physiological roles, but contributes to cancer progression in several ways. Angiogenesis increases vascular permeability to allow for inflammatory cells to accumulate, increases nutrient availability to developing tumor cells, and, later in cancer progression, provides an easily accessible avenue for apoptosis. The presence of reactive oxygen species (ROS) in tissues leads to DNA damage, genomic instability, and in a mouse model was shown to induce transformation of mouse mammary epithelial cells into malignant cells. MMP-3 activity has been shown to induce ROS via cleavage of the GTPase Rac1 at the cell surface, an event which is also likely to contribute to cancer initiation.[27] Several reports have demonstrated a role for MMPs affecting apoptosis. MMPs –1, 2, –3, –7, –9, and –11 have all been shown to have anti-apoptotic effects.[28] However, it should be noted that MMP-2, –3, –7, –9, and -11 have also been shown to have proapoptotic effects, depending on the circumstance.[28] The proapoptotic functions of MMP-7 are perhaps the most well studied; we now know that MMP-7 cleaves the Fas ligand at the cell surface, protecting cells from apoptosis, and thereby conferring chemoresistance.[29,30] During inflammation, the deposition of the Fas ligand cleavage products and many other MMP-activated signals may also extend immunosuppressive effects that supports tumor progression and ultimately drives metastasis.[31]

While chronic inflammation leading to cancer can occur as a result of internal disease, such as inflammatory bowel disease[32] or in autoimmune diseases such as arthritis or lupus,[33] the presence of certain chronic viral or bacterial infections or environmental assaults has been shown to potentiate these effects and further increases a patient's susceptibility to cancer. For example, chronic *Helicobacter pylori* infection, an infection associated with chronic heartburn, has been shown to increase MMP-3 and MMP-7 levels.[21,34] In fact,

H. pylori stimulated expression of several MMPs, which could be detected in the serum and were shown to be a viable option for predicting gastric cancer.[34] Chronic infection of the liver in patients with hepatitis B and C leads to development of cirrhotic nodules, leading to a progressive decrease in vasculature in the tissue and thus creating a hypoxic environment. Multiple studies demonstrate that hypoxia inducible factors lead to increased MMP expression.[35] Expression of these MMPs then causes local changes including angiogenesis and EMT, with these hypoxia-driven changes known to be a major stimulus driving hepatocellular carcinoma. Similarly, a recent study demonstrated for the first time that women who are coinfected with *Chlamydia trachomatis* and HPV exhibit increased MMP-9 levels with a concomitant decrease in expression of its endogenous inhibitor reversion-inducing-cysteine-rich protein with Kazal motifs (RECK). As evidenced by cervical smears, this imbalance was shown to have a role in driving cervical carcinogenesis as it was significantly associated with high-grade cervical diseases.[36] Cigarette smoke has been demonstrated to be another irritant which increases inflammation and thus MMP expression, setting the stage for cancer. An *in vitro* study provided direct evidence that the contents of cigarette smoke increased MMP-2 expression in normal oral fibroblasts and conferred migratory potential to the cells.[37] Another study demonstrates cigarette smoke activates NF-kB in a variety of cell types, which as discussed in preceding paragraphs, is an activator of expression of several MMPs.[38] MMPs are known to be upregulated in tissues of chronic smokers, and the link between smoking and a variety of cancers is well established.[39] As such, it can be concluded that the increase in MMP levels resulting from smoking plays a significant role in initiating and driving progression of these cancers.

While how exactly cancers are initiated remains somewhat murky, there is sufficient evidence to suggest that MMPs have a key role. Further, cancer is frequently associated with comorbidities, and appreciating the link between diseased tissue, MMP levels, and cancer progression can be a valuable tool for clinicians. Understanding the direct contributions of MMPs to each stage of cancer will be critical to diagnosing cancers earlier and developing targeted treatment approaches. Leveraging such knowledge will translate clinically to decreased morbidity and improved overall survival.

7.3 Polymorphisms

Cancer has long been considered a genetic disease. While the fact that genes are frequently mutated during cancer progression has been well described, the idea that the genotype with which a person is born with can indicate a predisposal to cancer is less well established. Epidemiologic studies have been conducted in depth only for a few such genes, with the breast cancer (*BRCA*) gene and its predisposition for breast cancer being the prime example. Such research is worth investing in, as an individual in possession of knowledge of their genotype may take chemopreventive measures to lower risk or more proactively monitor their physical well-being in order to catch cancer early. Such measures will not only greatly improve survival rates, but will dramatically decrease health-care expenditures.

Research with the purpose of identifying the genetic polymorphisms which may prime individuals for developing cancer remains largely in its infancy and is fraught with limitations. One of the largest challenges is obtaining a sufficient sample size to perform adequate analyses. To identify which somatic polymorphisms pose a risk, tumor genomes must not only be sequenced, but lifestyle choices must also be considered to fully understand the implications of a polymorphism. For example, cigarette smoke or environmental pollution may potentiate the risk associated with a particular polymorphism, whereas a healthy person with that same polymorphism living in a clean environment may not be at risk. Furthermore, the influence of bias is difficult to completely eradicate during analysis even if the most rigorous statistical analyses are utilized. Contributing to this bias, negative correlative data are generally published more slowly, if at all, than positive.

Notably, there is a massive number of genes to be analyzed, each of which may have multiple polymorphisms that can occur. Further complicating this is that as there are two alleles per gene, and in some cases only homozygotes may pose a predisposition, whereas in others a single allele, or a heterozygote, may be sufficient. The common nomenclature used to distinguish genotypes refers to a model in which, if we consider an allele which can be represented as "A" or "B", the risk occurs in the dominant model if the genotype is either AA or AB. In the recessive case, for a risk to be posed, the genotype of the individual must be BB.

There is an abundance of clinical studies in which genetic polymorphisms of different MMPs have been assessed for their correlation with cancer risk. As noted, however, it is often quite difficult for a single study to successfully establish a correlation with risk, largely owing to sample size. As a result, the published literature is often conflicting, with the same polymorphism sometimes being deemed not a risk by one study but yet at risk by another. For example, there is a known single-nucleotide polymorphism (SNP) in the MMP-2 promoter at −1306 bp relative to the transcription start site in which a C to T substitution occurs, abolishing the binding site of transcription factor Sp1 and can thus result in lower expression levels of MMP-2.[40,41] Several studies have demonstrated that the CC or CT polymorphisms correlate to increased risk of cancer development but do not correlate to metastasis.[42,43] However, another study suggests that neither genotype contributes to gastric cancer development but breast cancer patients with a CC genotype are at increased risk of lymphatic and venous invasion.[44] These data are also supported by an *in vitro* study of colon cancer demonstrating that this polymorphism is also associated with increased depth of tumor invasion.[45] In most studies conducted by a single research group on gene polymorphisms and their association with cancer, the number of patients assessed per genotype group will, on average, examine anywhere from 100 to 500 patients and typically will focus on a single cancer type. Between small sample sizes and regional and ethnic diversity, results of these studies may often conflict with one another. To best address these challenges and to reduce bias from small sample size, meta-analyses are frequently conducted. A meta-analysis is a statistical technique in which the findings from independent studies are combined, sorted, and reanalyzed all together. Such an approach allows for a more objective analysis of the data collected and should decrease false results.[46] Meta-analyses are particularly useful for establishing if a given polymorphism affects overall cancer risk or is only cancer-type specific. As different tissues tend to have different expression profiles for MMPs and transcription factors, a polymorphism may be relevant in one tissue but not in another. The benefit of meta-analyses is that the large sample size can allow for these differences to be teased apart and significance to be established.[47] For an in-depth review of how meta-analyses are conducted, please refer to the review by da Costa and Jüni.[46]

For the purpose of this review, we will focus primarily on meta-analyses in an effort to present the most thorough and least biased data regarding genetic variances of MMPs. When data on the MMP-2 –1306 C/T SNP are compiled, 26 comparisons were found and comprised 8297 cases and 10,566 controls. From such a large dataset, several different conclusions can be drawn. The –1306 TT and CT genotypes were deemed protective, as they are associated with lower cancer risk in head and neck, lung, gastric, and esophageal cancer subgroups. Subgroup analysis shows that smokers carrying the C > T genotype also have a decreased cancer risk. However, no association between MMP-2 –1306 C > T for either the dominant or recessive model for breast cancer or colorectal cancer was found.[47]

Polymorphisms in the MMP-9 gene are perhaps the most studied of any MMP to date. The reasons for this are two fold: first, MMP-9 is found to be expressed in nearly every cancer type, and is usually associated with accelerated disease progress and poorer survival rates.[48] Second, four different polymorphisms have been identified. One of these is a C/T polymorphism in the promoter at –1562 bp which is associated with higher transcriptional activity of the gene and thus has been linked to higher MMP-9 levels. Meta-analyses suggest that while there is no overall increased risk for cancer for individuals carrying this polymorphism,[43,49] cases which have genotype TT or CT are at increased risk of metastasis under the dominant model. However, no protective association was found between genotype TT and metastasis under the recessive model.[43] For the other three known MMP-9 polymorphisms, which include a C to G substitution in exon 10 (P574R, rs2250889), G to A substitution in exon 6 (R279Q, rs17576), and an intron polymorphism with a G to A substitution (rs3787268), no correlation with overall cancer risk was determined. Notably, subgroup analysis of rs3787268, however, indicated an increased risk for breast cancer in Native American women.[49,50]

The promoter region of MMP-1 contains another commonly known polymorphism. At −1607 bp, there exists a guanine insertion/deletion polymorphism, with the allele for this promoter having either a single guanine nucleotide (1G) or having two (2G). *In vitro* analysis has described a situation in which the 2G polymorphism (1G/2G or 2G/2G) creates a binding site for the ubiquitous transcription factor Ets. Cells with either a 1G/2G or 2G/2G were thus found to be more transcriptionally active with higher expression of MMP-1 than cells with 1G/1G genotype.[51] A meta-analysis of this promoter found that individuals carrying the 2G/2G phenotype have an increased risk for colorectal cancer, head and neck cancer, and renal cancer.[47] A separate study in Caucasians of 456 lung cancer cases and 451 controls found that the 2G/2G genotype is not in general indicative of a lung cancer risk. However, the study found that 2G/2G men who were current smokers had a threefold increase lung cancer risk compared to their nonsmoker counterparts. Furthermore, the study found that risk was still elevated in male patients who were formerly heavy smokers but had quit. This correlation was not significant though in light smokers. Interestingly, for all groups, men with the 2G phenotype were at a higher risk than women.[52]

Several polymorphisms in the MMP-12 gene have been described, the most well studied of which is SNP in the promoter region of MMP-12 at −82 bp in which there is an A to G substitution. It is believed this substitution affects transcriptional activity and leads to higher levels of MMP-12 expression. While this SNP is not reflective of an overall increased risk for cancer, two different meta-analyses have found that after subgroup stratification, there is a strong association of the G phenotype with ovarian cancer.[53,54] Considering that ovarian cancer remains one of the most deadly cancers for women due to late diagnosis and few treatment options, this knowledge could be leveraged by women to maintain vigilance in order to ensure the disease is caught early, leading to decreased morbidity and increased survival rates.

Polymorphisms in several other MMPs have been identified and investigated for their association with cancer risk. Several studies have investigated MMP-3 and MMP-7, although the data are conflicting and remain largely inconclusive. With cancer costs projected to reach $158 billion by 2020 for the United States alone, with roughly 500,000 people dying annually from cancer, better approaches to treatment are necessary. As most cancer cases can be cured if caught early enough before metastasis happens,[55] improving detection methods is critical to fight against cancer. Understanding who is at risk in order to closely monitor their health and catch the disease early on is a key piece to this puzzle.

7.4 Leveraging MMP Expression Changes as Biomarkers for Cancer

Because MMPs have been implicated in a number of pathological processes contributing to cancer progression, the latter including tumorigenesis, cancer cell invasion, metastasis, and dysregulated angiogenesis,[26,56–59] these proteases have potential for use as biomarkers for standard clinical oncology practice. For a summary of MMPs and their contribution to cancer progression and patient prognosis see Table 7.1. Biopsies remain the clinical standard for determining cancer grade and stage. In a recent study, MMP-1 and -9 protein expression in tumor-free mucosa was identified as critical prognostic factors for the prediction of cancer-specific survival (CSS) in colon cancer. This study relied on patient biopsies of the intestinal mucosa which was assessed for MMP levels using a standard enzyme-linked immunosorbent assay (ELISA).[60] In the case of lung cancer, MMP-9 levels were found to be higher in both non-small-cell lung cancer (NSCLC) and small-cell lung cancer (SCLC) when compared to normal surrounding tissues. In addition to these findings, this research group also found that MMP-9 expression correlated with tumor grade, size, and lymph node metastasis, which displayed an even more prominent expression relative to the primary

Table 7.1. Tissue specific expression of MMPs.

Matrix Metalloproteinase	Associated Cancer(s)
proMMP-1	Rectal carcinoma (serum)
MMP-1	Bladder (urine), colon (mucosa/blood)
MMP-2	Bladder (urine), breast (biopsy/serum), colon (mucosa/serum), glioma (biopsy/serum), prostate (urine)
MMP-3	Breast (biopsy), pancreatic (biopsy)
MMP-7	Colon (serum), rectal carcinoma (serum)
MMP-9	Bladder (urine), breast (biopsy/serum), colon (mucosa/serum), glioma (biopsy/serum), non-small-cell lung carcinoma (NSCLC) (biopsy/serum), prostate (urine), small-cell lung carcinoma (biopsy/serum)
MMP-10	Bladder (urine)
MMP-12	Skin (biopsy)
MMP-13	NSCLC (serum)
MT1-MMP (MMP-14)	Breast (biopsy), cervical carcinoma (biopsy), gastric (biopsy), nasopharyngeal carcinoma (biopsy), NSCLC (biopsy)
MT2-MMP (MMP-15)	Gastric (biopsy)
MMP-26	NSCLC (biopsy)

lesion.[61] Similar to the previously discussed MMPs, high membrane-type 1 MMP (MT1-MMP) expression has been causatively linked in gastric cancer progression and to a lesser extent linked to clinical stage, distant metastases, and differentiated degree of various other cancers such as NSCLC, cervical, and nasopharyngeal carcinomas.[62–67] Through the use of immunohistochemistry (IHC), elevated levels of MT1-MMP expression were also found in tumor biopsies of breast cancer patients relative to normal breast tissue. These findings were positively correlated with the tumor stage, lymph node metastasis, and overall metastatic potential in addition to being inversely correlated with patient survival.[68]

While analysis of tissue structure and the appearance of cells present in the tissue are frequently used to confirm the presence and determine the stage of cancer, they are limited to the small area of tissue excised. As such, performing additional laboratory tests on these samples to confirm the presence of biomarkers will decrease false positive results and improve reliability. Changes in the pathological conditions of the tumor microenvironment and individual cancer cells are often collectively reflected in bodily fluids. Many studies indicate that it is feasible to use measurements of MMP levels in these fluids as a promising, noninvasive tool for both diagnosing and monitoring the progression of various human cancers. A meta-analysis of MMP-7 expression levels in serum found that levels were increased in the serum of colorectal cancer patients relative to control subjects. This finding establishes MMP-7 as an independent biomarker for colorectal cancer.[61] Similarly, high preoperative MMP-2 serum levels were found to be associated with comparable patient survival rates as seen in patients with estrogen receptor (ER)-negative, higher histologic grade, or higher nuclear grade breast cancers.[69] In general, it has been found that patients with increased expression of any of the MMPs in serum have a higher risk of poor prognosis and tumor relapse in breast cancer.[70,71] Surprisingly, in lung cancer patients, MMP-9 levels have been found to be elevated in patient blood samples but not in bronchial lavage fluid. This study also found that MMP-9 can be used as a biomarker that can even differentiate between malignant and benign lung disease.[72,73]

Even less invasive, urine tests can also be adapted to assay for MMP expression. MMP-2 and -9, along with their dimerized counterparts, can be detected in the urine of patients. This suggests a possible role of high molecular weight (HMW) MMP species as potential biomarkers in prostate and bladder cancer patients.[74] To support this finding, another group similarly found that HMW gelatinase species were consistently detected in the urine of prostate and bladder carcinoma patients, therefore bolstering the potential role of MMP dimers in predicting patient prognosis and clinical outcome. Even more interesting, this group also showed that MMP-9 dimers were 1.5-fold more frequently found in the urine of bladder cancer patients relative to prostate cancer patients, which in general had twofold higher amounts of MMP-9 present in their urine. This suggests that the presence of MMP-9 and MMP-9 dimers in patient urine may serve as an important differential biomarker for identifying the presence of a tumor as well as its potential location.[75,76]

Although expression of MMPs is conventionally used for diagnostic and patient prognosis purposes, recent evidence has also suggested a potential role for MMPs in monitoring disease-free survival. In a recent study investigating proMMP-1, MMP-2, -7, and -9 in the serum of patients with rectal cancer, it was found that the pretreatment concentration of proMMP-1 may be clinically relevant when evaluating tumor mass. In addition to this, MMP-7 was also found as a potential prognostic factor for disease-free survival in men with rectal cancer without distant metastasis.[77]

Along similar lines, the presence of MMPs extends beyond tumor stage/grade and patient survival. In recent studies, MMPs have also been important in preliminarily assessing a patient's response to therapy. For example, fibroblast growth factor receptor 3 (FGFR3) is essential for bladder cancer progression. Due to this, many therapies have been invested into targeting this receptor in bladder cancer. Recently, a group has described the use of a monoclonal antibody (R3Mab), which specifically targets this receptor and have identified MMP-1 and pro-MMP-10 as potential biomarkers for effective R3Mab treatment in this subgroup of cancer patients. Inhibition of FGFR3 signaling through R3Mab treatment has been found to reduce the overall expression and secretion of MMP-1 and pro-MMP-10 in tumor tissues as well as mouse serum in addition to retarding further tumor development. Collectively, this group demonstrated for the first time that MMPs can be used as a positive read-out for monitoring the effect of R3Mab treatment in bladder cancer patients.[78]

7.5 Detecting and Monitoring Cancer via MMPs

Between the accumulation of various MMPs in the microenvironment of nearly all cancer types (refer to Table 7.1) and their use as biomarkers of cancer stage or metastasis, MMPs are promising for use in cancer diagnosis and/or monitoring cancer progression.

Multiple studies which use near-infrared (NIR) fluorescent or radiolabeling of MMP-specific probes have been conducted for the purpose of improved sensitivity for tumor imaging. Such approaches are noninvasive and thus are attractive for clinical applications. MMPSense probes are commercially available MMP imaging agents which have been used *in vivo* to assess MMP activity. The probes consist of MMP-cleavable peptides conjugated to NIR fluorescent moieties. Prior to MMP-mediated cleavage, the fluorescent signal is quenched and it is only upon cleavage of the peptide that the fluorophores decouple and a signal is produced.[79] Several mouse studies have been conducted which clearly demonstrate feasibility of these probes. Results have also demonstrated that these probes can allow for earlier detection of cancer, improved methods for monitoring cancer stage and progression, and opportunities to improve therapeutic approaches, including use in guided surgery.[80–82] Several other fluorogenic probes have been designed for use in imaging MMPs that incorporate a similar approach. These studies have tested the use of different fluorophores and/or substrates in an effort to reduce background signal and improve selectivity. Other fluorogenic approaches to identify MMP activity used substrates designed to mimic the cleavage sites of triple helical collagen. These probes demonstrated increased specificity for MMP-2 and MMP-9 with the signals showing higher tumor to muscle tissue ratios.[83,84]

While use of fluorescent MMP-specific probes have proven useful, nuclear approaches may be preferred for imaging of tissues deep in the body due to a higher signal strength. Additionally, they can be used simultaneously with electron emitter cancer therapies.[85] Using an approach similar to the cleavable fluorescent probes, radiolabeled probes have been developed. One study involved a system in which indium-111 was incorporated into an MMP-2 cleavable substrate. The radiolabeled region of the substrate was also conjugated to a membrane-anchoring double alkyl chain so that the system would remain associated with or near MMP-2-expressing cells. The presence of polyethylene glycol (PEG) confers hydrophilicity to delay clearance from tissues *in vivo*. Mouse studies demonstrated efficient clearance of the radioprobes from the blood and accumulation in tumors and indicated their superiority to NIR labels for tumor-imaging purposes.[85]

Blood tests are a standard practice for evaluating a patient's health and detecting changes that may indicate the onset of pathologies, including cancer. Implementing tests for the presence of or changes in the levels of MMPs could be a cost-efficient method which would not add any additional physical burden to patients. Making this standard practice for certain cancers or for select populations at risk of certain cancers could potentially be groundbreaking in terms of early detection and reducing morbidity. For example, ovarian cancer is rarely detected until patients reach an advanced stage and because of this the five-year survival rate for women affected remains at only 45%. The current approaches used for diagnosis have relatively poor sensitivity and specificity for early stages of ovarian carcinomas.[86] A clinical study evaluated a panel of biomarkers in serum, including MMP-7, using a combination of commercially available tests and ELISA-based assays. Results from this study demonstrated improved specificity and high sensitivity values for detection of early stage ovarian cancer, with values even higher for late stage cancer. Particularly noteworthy, a subset of biomarkers which also included MMP-7 were able to predict recurrence earlier than the currently used marker, cancer antigen 125 (CA125). Furthermore, this panel was able to detect residual disease in approximately half of asymptomatic patients who have positive results in their follow up laparotomy following their initial surgery; this was true for 96% of the patients when using the CA125 biomarker only.[87]

While blood samples are a readily available source of biomarkers, being able to detect the presence of or changes in cancer through urine would be even less invasive. The gold standard for bladder cancer detection and monitoring is through the use of cystoscopy and cytology. Although this method is sensitive, it is highly invasive and the procedure is estimated to cause urinary tract infections in as many as 16% of patients.[88] In one study, a biochip microarray was developed for detection of bladder cancer which used a modified fluorescent immunoassay approach. This biochip prototype was demonstrated to be capable of simultaneously measuring levels of five different bladder cancer biomarkers, two of which are MMP-7 and MMP-9.[89] As this chip relies on a urine sample for detection, this noninvasive approach holds potential to become the preferred method of early diagnosis. The use of multiple biomarkers to evaluate the presence/progression of cancer instead of a single biomarker will likely yield improved sensitivity while reducing false-positive results. Although this approach was proof-of-concept only, clinical applications are appreciable.

7.6 MMPs as Therapeutic Targets: Looking to the Future

Nearly all cancers overexpress MMPs, as these proteases are pleiotropic and contribute in a multitude of ways to all aspect of cancer progression, from tumor growth, to angiogenesis, to metastasis.[90] For this reason, pharmaceutical companies in the nineties invested millions of dollars into small-molecule inhibitors which bound in the active site of the protease. Ultimately though, these investments fell flat and the only MMP inhibitor to date approved for patient use is Periostat® (doxycycline hyclate), which is indicated for periodontal disease.[91] The elucidation of the complexity, diversity, and three-dimensional properties of MMPs is still being explored, and the knowledge of the basic science of MMPs can be leveraged to improve drug design.

Doxycycline hyclate is a chemically modified tetracycline used at subantimicrobial doses as a broad-spectrum MMP inhibitor. Due to its well-described pharmacokinetic/pharmacodynamic (PK/PD) profile and good safety record, doxycycline is currently being clinically tested in combination standard chemotherapeutics.[91] The main reason of failure for small-molecule inhibitors seems to be largely due to the poor selectivity of the drugs, leading to off-target effects which caused dose-limiting musculoskeletal adverse effects (i.e., debilitating joint pain).[92] As monoclonal antibodies (mAbs) are known to demonstrate higher selectivity, they are often preferred to traditional approaches to reduce off-target effects. An MMP-9-selective mAb currently is in clinical trials and is in the pipeline of Gilead Sciences.

GS-5745 is a humanized mAb which allosterically targets and noncompetitively inhibits MMP-9. Studies indicate that the epitope of this mAb include residues located around the calcium ion binding pocket.[93] Phase I trials preliminarily indicate an acceptable safety profile for this drug, both alone and in combination with chemotherapy.[94] A phase III clinical trial is currently recruiting for GS-5745 (NCT02545504) in combination with modified fluorouracil (5-FU), leucovorin (LV), and oxaliplatin (OXA) (mFOLFOX6) as a first-line treatment in patients with advanced gastric or gastroesophageal adenocarcinoma. A phase I trial is also in progress for solid tumors, and this drug is in trials for several other, noncancer applications including ulcerative colitis (phase III), Crohn's disease (phase II), chronic obstructive pulmonary disease (COPD) (phase I), and rheumatoid arthritis (phase I). A potential drawback with antibodies, however, is that they may be subject to rapid proteolysis, leading to their clearance from circulation. Additionally, therapeutic antibodies are administered parenterally, which is a burden on the patient.[95]

Extensive literature exists in which preclinical studies target various MMPs using a variety of approaches. Current research available on small-molecule inhibitors seems to focus largely on those that bind allosterically in an effort to reduce the off target effects observed previously. A druggable pocket of the MMP-14 hemopexin domain has been identified which should block MMP-14-involved protein–protein interactions. As MMP-14 is a membrane-bound protease capable of homo- and heterodimerizing via its hemopexin domain to drive cell invasion and metastasis,[96] disruption of such processes is critical to patient care. Further, the hemopexin domains of MMPs are not highly conserved,[97] suggesting that targeting this region can improve selectivity. Identification of a small-molecule inhibitor binding in this druggable pocket performed successfully *in vitro* to decrease cell migration and reduced tumor burden *in vivo*.[98]

The majority of the recent publications targeting MMPs involve a multi-approach system. At least two systems have been designed in which a cytotoxic agent is loaded into a carrier whose surface is studded with MMP-binding partners. In these cases, one of which is a liposome and the other a nanoparticle, selective interaction with

MMPs within the tumor environment leads to release of the cytotoxic agent inside or nearby tumor cells, leading to cancer cell death. These innovative ideas have been successful *in vitro*, and it will be interesting to see if *in vivo* data can recapitulate these promising results.[99,100]

7.7 Summary

The list of functions of MMPs in cancer is as long as it is diverse. From activation of signaling molecules to driving angiogenesis and metastasis, MMPs have been well characterized in how they contribute to tumorigenesis. But while there is extensive research into the basic science of MMPs, a clinical appreciation lags behind. Different MMPs contribute to tumorigenesis in a number of ways, and changes in their expression levels can be leveraged to improve detection of and monitoring progression of cancer. In order to decrease healthcare expenditures and improve patient outlook, methods for early detection of cancer are critical. Furthermore, as techniques to monitor changes in tumor progression and the tumor microenvironment improve, therapeutic intervention strategies will improve in turn. As MMPs are abundantly expressed in cancer and directly drive its progression, they remain attractive and viable targets for improving cancer management strategies.

References

1. Karin M and FR Greten (2005). NF-κB: linking inflammation and immunity to cancer development and progression. *Nature Reviews Immunology*, 5(10), 749–759.
2. Sternlicht, MD and Z Werb (2001). How matrix metalloproteinases regulate cell behavior. *Annual Review of Cell and Developmental Biology*, 17(1), 463–516.
3. Egeblad, M and Z Werb (2002). New functions for the matrix metalloproteinases in cancer progression. *Nature Reviews Cancer*, 2(3), 161–174.
4. Sternlicht, MD, A Lochter, CJ Sympson, B Huey, J-P Rougier, JW Gray, D Pinkel, MJ Bissell and Z Werb (1999). The stromal

proteinase MMP3/stromelysin-1 promotes mammary carcinogenesis. *Cell*, 98(2), 137–146.

5. Ha, HY, HB Moon, MS Nam, JW Lee, ZY Ryoo, TH Lee, KK Lee, BJ So, H Sato, M Seiki and DY Yu (2001). Overexpression of membrane-type matrix metalloproteinase-1 gene induces mammary gland abnormalities and adenocarcinoma in transgenic mice. *Cancer Research*, 61(3), 984–990.
6. Coussens, LM, CL Tinkle, D Hanahan and Z Werb (2000). MMP-9 supplied by bone marrow–derived cells contributes to skin carcinogenesis. *Cell*, 103(3), 481–490.
7. Jacobs, TW, C Byrne, G Colditz, JL Connolly and SJ Schnitt (1999). Radial scars in benign breast-biopsy specimens and the risk of breast cancer. *New England Journal of Medicine*, 340(6), 430–436.
8. Dragsbæk, K, JS Neergaard, HB Hansen, I Byrjalsen, P Alexandersen, SN Kehlet, AC Bay-Jensen, C Christiansen and MAKarsdal (2015). Matrix metalloproteinase mediated type I collagen degradation — an independent risk factor for mortality in women. *EBioMedicine*, 2(7), 723–729.
9. Bourboulia, D and WG Stetler-Stevenson (2010). Matrix metalloproteinases (MMPs) and tissue inhibitors of metalloproteinases (TIMPs): positive and negative regulators in tumor cell adhesion. *Seminars in Cancer Biology*, 20(3), 161–168.
10. Thomasset, N, A Lochter, CJ Sympson, LR Lund, DR Williams, O Behrendtsen, Z Werb and MJ Bissell (1998). Expression of autoactivated stromelysin-1 in mammary glands of transgenic mice leads to a reactive stroma during early development. *American Journal of Pathology*, 153(2), 457–467.
11. Coussens, LM and Z Werb (2002). Inflammation and cancer. *Nature*, 420(6917), 860–867.
12. Giraudo, E, M Inoue and D Hanahan (2004). An amino-bisphosphonate targets MMP-9–expressing macrophages and angiogenesis to impair cervical carcinogenesis. *Journal of Clinical Investigation*, 114(5), 623–633.
13. Han, YP, TL Tuan, H Wu, M Hughes and WL Garner (2001). TNF-alpha stimulates activation of pro-MMP2 in human skin through NF-(kappa)B mediated induction of MT1-MMP. *Journal of Cell Science*, 114(1), 131–139.
14. Cheng, JCH, CH Chou, ML Kuo and CY Hsieh (2006). Radiation-enhanced hepatocellular carcinoma cell invasion with MMP-9

expression through PI3K//Akt//NF-kappaB signal transduction pathway. *Oncogene*, 25(53), 7009–7018.
15. Kundu, JK and Y-J Surh (2008). Inflammation: gearing the journey to cancer. *Mutation Research/Reviews in Mutation Research*, 659(1–2), 15–30.
16. Yokoo, T and M Kitamura (1996). Dual regulation of IL-1 beta-mediated matrix metalloproteinase-9 expression in mesangial cells by NF-kappa B and AP-1. *American Journal of Physiology-Renal Physiology*, 270(1), F123–F130.
17. Li, J, GK-K Lau, L Chen, Dong S-S, Lan H-Y, Huang X-R, Y Li, JM Luk, YF Yuan and XY Guan (2011). Interleukin 17A promotes hepatocellular carcinoma metastasis via NF-kB induced matrix metalloproteinases 2 and 9 expression. *PLoS One*, 6(7), e21816.
18. Liacini, A, J Sylvester, Li W Qing, W Huang, F Dehnade, M Ahmad and M Zafarullah (2003). Induction of matrix metalloproteinase-13 gene expression by TNF-α is mediated by MAP kinases, AP-1, and NF-κB transcription factors in articular chondrocytes. *Experimental Cell Research*, 288(1), 208–217.
19. Liacini, A, J Sylvester, WQ Li and M Zafarullah (2002). Inhibition of interleukin-1-stimulated MAP kinases, activating protein-1 (AP-1) and nuclear factor kappa B (NF-kappa B) transcription factors down-regulates matrix metalloproteinase gene expression in articular chondrocytes. *Matrix Biology*, 21(3), 251–262.
20. Vincenti, MP, CI Coon and CE Brinckerhoff (1998). Nuclear factor kappa B/p50 activates an element in the distal matrix metalloproteinase 1 promoter in interleukin-1β-stimulated synovial fibroblasts. *Arthritis & Rheumatism*, 41(11), 1987–1994.
21. Wroblewski, LE, P-JM Noble, A Pagliocca, DM Pritchard, CA Hart, F Campbell, AR Dodson, GJ Dockray and A Varro (2003). Stimulation of MMP-7 (matrilysin) by helicobacter pylori in human gastric epithelial cells: role in epithelial cell migration. *Journal of Cell Science*, 116(14), 3017–3026.
22. Dufour, A and CM Overall (2013). Missing the target: matrix metalloproteinase antitargets in inflammation and cancer. *Trends in Pharmacological Sciences*, 34(4), 233–242.
23. Whitelock, JM, AD Murdoch, RV Iozzo and PA Underwood (1996). The degradation of human endothelial cell-derived perlecan and release of bound basic fibroblast growth factor by stromelysin, collagenase,

plasmin, and heparanases. *Journal of Biological Chemistry*, 271(17), 10079–10086.

24. McCawley, LJ and LM Matrisian (2001). Matrix metalloproteinases: they're not just for matrix anymore! *Current Opinion in Cell Biology*, 13(5), 534–540.
25. Kazushi, I, A Hiramatsu, D Fukushima, MD Pierschbacher and Y Okada (1997). Degradation of decorin by matrix metalloproteinases: identification of the cleavage sites, kinetic analyses and transforming growth factor-β1 release. *Biochemical Journal*, 322(3), 809–814.
26. Lee, S, SM Jilani, GV Nikolova, D Carpizo and ML Iruela-Arispe (2005). Processing of VEGF-A by matrix metalloproteinases regulates bioavailability and vascular patterning in tumors. *Journal of Cell Biology*, 169(4), 681–691.
27. Radisky, DC, DD Levy, LE Littlepage, H Liu, CM Nelson, JE Fata, Leake D, EL Godden, DG Albertson, MA Nieto, Z Werb and MJ Bissell (2005). Rac1b and reactive oxygen species mediate MMP-3-induced EMT and genomic instability. *Nature*, 436(7047), 123–127.
28. Mannello, F, F Luchetti, E Falcieri and S Papa (2005). Multiple roles of matrix metalloproteinases during apoptosis. *Apoptosis*, 10(1), 19–24.
29. Mitsiades, N, W-h Yu, V Poulaki, M Tsokos and I Stamenkovic (2001). Matrix metalloproteinase-7-mediated cleavage of fas ligand protects tumor cells from chemotherapeutic drug cytotoxicity. *Cancer Research*, 61(2), 577–581.
30. Powell, WC, B Fingleton, CL Wilson, M Boothby and LM Matrisian (1999). The metalloproteinase matrilysin proteolytically generates active soluble fas ligand and potentiates epithelial cell apoptosis. *Current Biology*, 9(24), 1441–1447.
31. Kim, R, M Emi, K Tanabe and K Arihiro (2006). Tumor-Driven evolution of immunosuppressive networks during malignant progression. *Cancer Research*, 66(11), 5527–5536.
32. Mantovani, A, P Allavena, A Sica and F Balkwill (2008). Cancer-related inflammation. *Nature*, 454(7203), 436–444.
33. Smedby, KE, H Hjalgrim, J Askling, ET Chang, H Gregersen, A Porwit-MacDonald, C Sundström, M Akerman, M Melbye, B Glimelius and HO Adami (2006). Autoimmune and chronic inflammatory disorders and risk of non-Hodgkin lymphoma by subtype. *Journal of the National Cancer Institute*, 98(1), 51–60.

34. Yeh, Y-C, B-S Sheu, H-C Cheng, Y-L Wang, H-B Yang and J-J Wu (2010). Elevated serum matrix metalloproteinase-3 and-7 in H. pylori-related gastric cancer can be biomarkers correlating with a poor survival. *Digestive diseases and sciences*, 55(6), 1649–1657.
35. Arzumanyan, A, HM Reis and MA Feitelson (2013). Pathogenic mechanisms in HBV-and HCV-associated hepatocellular carcinoma. *Nature Reviews Cancer*, 13(2), 123–135.
36. Discacciati, MG, F Gimenes, PC Pennacchi, F Faião-Flores, LC Zeferino, SM Derchain, JCTeixeira, MC Costa, M Zonta, L Termini, E Boccardo, A Longatto-Filho, ME Consolaro, LL Villa and SS Maria-Engler (2015). MMP-9/RECK Imbalance: a mechanism associated with High-Grade cervical lesions and genital infection by human papillomavirus and chlamydia trachomatis. *Cancer Epidemiology Biomarkers & Prevention*, 24(10), 1539–1547.
37. Pal, A, G Melling, EE Hinsley, TD Kabir, HE Colley, C Murdoch and DW Lambert (2013). Cigarette smoke condensate promotes pro-tumourigenic stromal–epithelial interactions by suppressing miR-145. *Journal of Oral Pathology & Medicine*, 42(4), 309–314.
38. Shishodia, S, P Potdar, CG Gairola and BB Aggarwal (2003). Curcumin (diferuloylmethane) down-regulates cigarette smoke-induced NF-kappaB activation through inhibition of IκBα kinase in human lung epithelial cells: correlation with suppression of COX-2, MMP-9 and cyclin D1. *Carcinogenesis*, 24(7), 1269–1279.
39. Lagente, V and E Boichot (2010). Role of matrix metalloproteinases in the inflammatory process of respiratory diseases. *Journal of Molecular and Cellular Cardiology*, 48(3), 440–444.
40. Dey, S, N Ghosh, D Saha, K Kesh, A Gupta and S Swarnakar. Matrix metalloproteinase-1 (MMP-1) promoter polymorphisms are well linked with lower stomach tumor formation in eastern Indian population. *PLoS One,* 9(2), e88040.
41. Yu, C, Y Zhou, X Miao and P Xiong, W Tan and D Lin (2004). Functional haplotypes in the promoter of matrix metalloproteinase-2 predict risk of the occurrence and metastasis of esophageal cancer. *Cancer Research*, 64(20), 7622–7628.
42. Miao, X, C Yu, W Tan, P Xiong, G Liang, W Lu and Lin D (2003). A functional polymorphism in the matrix metalloproteinase-2 gene promoter (–1306C/T) is associated with risk of development but not metastasis of gastric cardia adenocarcinoma. *Cancer Research*, 63(14), 3987–3990.

43. Liu, D, H Guo, Y Li, X Xu, K Yang and Y Bai (2012). Association between polymorphisms in the promoter regions of matrix metalloproteinases (MMPs) and risk of cancer metastasis: a meta-analysis. *PLoS One*, 7(2), e31251.
44. Wu, C-Y, M-S Wu, Y-J Chen, C-J Chen, H-P Chen, C-T Shun, GH Chen, SP Huang and JT Lin (2007). Clinicopathological significance of MMP-2 and TIMP-2 genotypes in gastric cancer. *European Journal of Cancer*, 43(4), 799–808.
45. Xu, E, M Lai, B Lv, X Xing, Q Huang and X Xia (2004). A single nucleotide polymorphism in the matrix metalloproteinase-2 promoter is associated with colorectal cancer. *Biochemical and Biophysical Research Communications*, 324(3), 999–1003.
46. da Costa, BR and P Jüni (2014). Systematic reviews and meta-analyses of randomized trials: principles and pitfalls. *European Heart Journal*, 35(47), 3336–3345.
47. Peng, B, L Cao, X Ma, W Wang, D Wang and L Yu (2010). Meta-analysis of association between matrix metalloproteinases 2, 7 and 9 promoter polymorphisms and cancer risk. *Mutagenesis*, 25(4), 371–379.
48. Hadler-Olsen, E, J-O Winberg and L Uhlin-Hansen (2013). Matrix metalloproteinases in cancer: their value as diagnostic and prognostic markers and therapeutic targets. *Tumor Biology*, 34(4), 2041–2051.
49. Zhang, X, G Jin, J Li and L Zhang (2015). Association between four MMP-9 polymorphisms and breast cancer risk: a meta-analysis. *Medical Science Monitor: International Medical Journal of Experimental and Clinical Research*, 21, 1115.
50. Slattery, ML, E John, G Torres-Mejia, M Stern, A Lundgreen, L Hines, A Giuliano, K Baumgartner, J Herrick and RK Wolff (2013). Matrix metalloproteinase genes are associated with breast cancer risk and survival: the breast cancer health disparities study. *PLoS One*, 8(5), e63165.
51. Rutter, JL, TI Mitchell, G Butticè, J Meyers, JF Gusella, LJ Ozelius and CE Brinckerhoff (1998). A single nucleotide polymorphism in the matrix metalloproteinase-1 promoter creates an Ets binding site and augments transcription. *Cancer Research*, 58(23), 5321–5325.
52. Zhu, Y, MR Spitz, L Lei, GB Mills and X Wu (2001). A single nucleotide polymorphism in the matrix metalloproteinase-1 promoter enhances lung cancer susceptibility. *Cancer Research*, 61(21), 7825–7829.

53. Liu, L, J Sun, G Li, B Gu, X Wang, H Chi and F Guo (2015). Association between MMP-12-82A/G polymorphism and cancer risk: a meta-analysis. *International Journal of Clinical and Experimental Medicine*, 8(8), 11896.
54. Chen, S-S, J Song, X-Y Tu, J-H Zhao and X-Q Ye (2015). The association between MMP-12 82 A/G polymorphism and susceptibility to various malignant tumors: a meta-analysis. *International Journal of Clinical and Experimental Medicine [Internet]*, 8(7), 10845–10854. http://europepmc.org/abstract/MED/26379878; http://europepmc.org/articles/PMC4565261?pdf=render; http://europepmc.org/articles/PMC4565261.
55. Chambers, AF, AC Groom and IC MacDonald (2002). Metastasis: dissemination and growth of cancer cells in metastatic sites. *Nature Reviews Cancer*, 2(8), 563–572.
56. Kessenbrock, K, V Plaks and Z Werb (2010). Matrix metalloproteinases: regulators of the tumor microenvironment. *Cell*, 141(1), 52–67.
57. Curran, S and GI Murray (1999). Matrix metalloproteinases in tumour invasion and metastasis. *Journal of Pathology*, 189(3), 300–308.
58. Curran, S and GI Murray (2000). Matrix metalloproteinases: molecular aspects of their roles in tumour invasion and metastasis. *European Journal of Cancer*, 36(13 Spec No), 1621–1630.
59. Yu, Q and I Stamenkovic (2000). Cell surface-localized matrix metalloproteinase-9 proteolytically activates TGF-beta and promotes tumor invasion and angiogenesis. *Genes & Development*, 14(2), 163–176.
60. Langenskiold, M, ML Ivarsson, L Holmdahl, P Falk, C Kabjorn-Gustafsson and E Angenete (2013). Intestinal mucosal MMP-1 — a prognostic factor in colon cancer. *Scandinavian Journal of Gastroenterology*, 48(5), 563–569.
61. Xing, XJ, XH Gu and TF Ma (2014). Relationship of serum MMP-7 levels for colorectal cancer: a meta-analysis. *Tumour Biology: Journal of the International Society for Oncodevelopmental Biology and Medicine*, 35(10), 10515–10522.
62. Caenazzo, C, M Onisto, L Sartor, R Scalerta, A Giraldo, D Nitti and S Garbisa (1998). Augmented membrane type 1 matrix metalloproteinase (MT1–MMP): MMP-2 messenger RNA ratio in gastric carcinomas with poor prognosis. *Clinical Cancer Research: An Official Journal of the American Association for Cancer Research*. 4(9), 2179–2186.

63. Mori, M, K Mimori, T Shiraishi, T Fujie, K Baba, H Kusumoto, M Haraguchi, H Ueo and T Akiyoshi (1997). Analysis of MT1–MMP and MMP2 expression in human gastric cancers. *International Journal of Cancer*, 74(3), 316–321.
64. Dong, Y, G Chen, M Gao and X Tian (2015). Increased expression of MMP14 correlates with the poor prognosis of Chinese patients with gastric cancer. *Gene*, 563(1), 29–34.
65. Wang, H, X Zhang, L Huang, J Li, S Qu and F Pan (2014). Matrix metalloproteinase-14 expression and its prognostic value in cervical carcinoma. *Cell Biochemistry and Biophysics*, 70(2), 729–734.
66. Wang, YZ, KP Wu, AB Wu, ZC Yang, JM Li, YL Mo, M Xu, B Wu and ZX Yang (2014). MMP-14 overexpression correlates with poor prognosis in non-small cell lung cancer. *Tumour Biology: Journal of the International Society for Oncodevelopmental Biology and Medicine*, 35(10), 9815–9821.
67. Yan, TH, ZH Lin, JH Jiang, SW Lu, HX Que, MA Chen, XS He, GB Que, JN Xiao and YQ Chen (2015). Matrix metalloproteinase 14 overexpression is correlated with the progression and poor prognosis of nasopharyngeal carcinoma. *Archives of Medical Research*, 46(3), 186–192.
68. Li, Y, G Cai, S Yuan, Y Jun, N Li, L Wang, F Chen, R Ling and J Yun (2015). The overexpression membrane type 1 matrix metalloproteinase is associated with the progression and prognosis in breast cancer. *American Journal of Translational Research*, 7(1), 120–127.
69. Song, N, H Sung, JY Choi, S Han, S Jeon, M Song, Y Lee, C Park, SK Park, KM Lee, KY Yoo, DY Noh, SH Ahn, SA Lee and D Kang (2012). Preoperative serum levels of matrix metalloproteinase-2 (MMP-2) and survival of breast cancer among Korean women. *Cancer Epidemiology Biomarkers & Prevention: A Publication of the American Association for Cancer Research, Cosponsored by the American Society of Preventive Oncology*, 21(8), 1371–1380.
70. Ren, F, R Tang, X Zhang, WM Madushi, D Luo, Y Dang, Z Li, K Wei and G Chen (2015). Overexpression of MMP family members functions as prognostic biomarker for breast cancer patients: a systematic review and meta-analysis. *PLoS One*, 10(8), e0135544.
71. Somiari, SB, CD Shriver, C Heckman, C Olsen, H Hu, R Jordan, Arciero C, S Russell, G Garguilo, J Hooke and RI Somiari (2006). Plasma concentration and activity of matrix metalloproteinase 2 and 9

in patients with breast disease, breast cancer and at risk of developing breast cancer. *Cancer Letters*, 233(1), 98–107.

72. Koc, M, D Ediger, F Budak, M Karadag, HB Oral, E Uzaslan, E Ege and RO Gözü (2006). Matrix metalloproteinase-9 (MMP-9) elevated in serum but not in bronchial lavage fluid in patients with lung cancer. *Tumori*, 92(2), 149–154.
73. El-Badrawy, MK, AM Yousef, D Shaalan and AZ Elsamanoudy (2014). Matrix metalloproteinase-9 expression in lung cancer patients and its relation to serum mmp-9 activity, pathologic type, and prognosis. *Journal of Bronchology & Interventional Pulmonology*, 21(4), 327–334.
74. Mohammed, MA, MF Seleim, MS Abdalla, HM Sharada and AH Abdel Wahab (2013). Urinary high molecular weight matrix metalloproteinases as non-invasive biomarker for detection of bladder cancer. *BMC Urology*, 13, 25.
75. Roy, R, G Louis, KR Loughlin, D Wiederschain, SM Kilroy, CC Lamb, D Zurakowski and MA Moses (2008). Tumor-specific urinary matrix metalloproteinase fingerprinting: identification of high molecular weight urinary matrix metalloproteinase species. *Clinical Cancer Research: An Official Journal of the American Association for Cancer Research*, 14(20), 6610–6617.
76. Eissa, S, R Ali-Labib, M Swellam, M Bassiony, F Tash, and TM El-Zayat (2007). Noninvasive diagnosis of bladder cancer by detection of matrix metalloproteinases (MMP-2 and MMP-9) and their inhibitor (TIMP-2) in urine. *European Urology*, 52(5), 1388–1396.
77. Fuksiewicz, M, B Kotowicz, A Rutkowski and M Kowalska (2015). The matrix metalloproteinase-7 and pro-enzyme of metalloproteinase-1 as a potential marker for patients with rectal cancer without distant metastasis. *Tumour Biology: Journal of the International Society for Oncodevelopmental Biology and Medicine*, 36(5), 3629–3635.
78. Du, X, BC Lin, QR Wang, H Li, E Ingalla, J Tien, I Rooney, A Ashkenazi, E Penuel and J Qing (2014). MMP-1 and Pro-MMP-10 as potential urinary pharmacodynamic biomarkers of FGFR3-targeted therapy in patients with bladder cancer. *Clinical Cancer Research: An Official Journal of the American Association for Cancer Research*, 20(24), 6324–6335.

79. Fields, GB and MJ Stawikowski (2016). Imaging matrix metalloproteinase activity implicated in breast cancer progression. *Breast Cancer: Methods and Protocols,* 1406, 303–329.
80. Clapper, ML, HH Hensley, W-CL Chang, K Devarajan, MT Nguyen and HS Cooper (2011). Detection of colorectal adenomas using a bioactivatable probe specific for matrix metalloproteinase activity. *Neoplasia*, 13(8), 685–691.
81. Xie, B-W, IM Mol, S Keereweer, ER Van Beek, I Que, TJ Snoeks, A Chan, EL Kaijzel and CW Löwik (2012). Dual-wavelength imaging of tumor progression by activatable and targeting near-infrared fluorescent probes in a bioluminescent breast cancer model. *PLoS One*, 7(2), e31875.
82. Zhang, Q, M Yang, J Shen, LM Gerhold, RM Hoffman and HR Xing (2010). The role of the intravascular microenvironment in spontaneous metastasis development. *International Journal of Cancer*, 126(11), 2534–2541.
83. Akers, WJ, B Xu, H Lee, GP Sudlow, GB Fields, S Achilefu and WB Edwards (2012). Detection of MMP-2 and MMP-9 activity in vivo with a triple-helical peptide optical probe. *Bioconjugate Chemistry*, 23(3), 656–663.
84. Zhang, X, J Bresee, GB Fields and WB Edwards (2014). Near-infrared triple-helical peptide with quenched fluorophores for optical imaging of MMP-2 and MMP-9 proteolytic activity in vivo. *Bioorganic & Medicinal Chemistry Letters*, 24(16), 3786–3790.
85. Temma, T, H Hanaoka, A Yonezawa, N Kondo, K Sano, T Sakamoto, M Seiki, M Ono and H Saji (2014). Investigation of a MMP-2 activity-dependent anchoring probe for nuclear imaging of cancer. *PLoS One*, 9(7), e102180.
86. Vaughan, S, JI Coward, RC Bast, A Berchuck, JS Berek, JD Brenton, G Coukos, CC Crum, R Drapkin, D Etemadmoghadam, M Friedlander, H Gabra, SB Kaye, CJ Lord, E Lengyel, DA Levine, IA McNeish, U Menon, GB Mills, KP Nephew, AM Oza, AK Sood, EA Stronach, H Walczak, DD Bowtell and FR Balkwill (2011). Rethinking ovarian cancer: recommendations for improving outcomes. *Nature Reviews Cancer*, 11(10), 719–725.
87. Havrilesky, LJ, CM Whitehead, JM Rubatt, RL Cheek, J Groelke, Q He, DP Malinowski, TJ Fischer and A Berchuck (2008). Evaluation of biomarker panels for early stage ovarian cancer detection and

monitoring for disease recurrence. *Gynecologic Oncology*, 110(3), 374–382.

88. Gkialas, I, G Papadopoulos, L Iordanidou, G Stathouros, C Tzavara, A Gregorakis and M Lykourinas (2008). Evaluation of urine tumor-associated trypsin inhibitor, CYFRA 21-1, and urinary bladder cancer antigen for detection of high-grade bladder carcinoma. *Urology*, 72(5), 1159–1163.
89. Gogalic, S, U Sauer, Doppler S and C Preininger (2015). Bladder cancer biomarker array to detect aberrant levels of proteins in urine. *Analyst*, 140(3), 724–735.
90. Overall, CM and RA Dean (2006). Degradomics: systems biology of the protease web. Pleiotropic roles of MMPs in cancer. *Cancer and Metastasis Reviews*, 25(1), 69–75.
91. Cathcart, J, A Pulkoski-Gross and J Cao (2015). Targeting matrix metalloproteinases in cancer: bringing new life to old ideas. *Genes & Diseases*, 2(1), 26–34.
92. Zucker, S, J Cao and W-T Chen (2000). Critical appraisal of the use of matrix metalloproteinase inhibitors in cancer treatment. *Oncogene*, 19(56), 6642–6650.
93. McCauley, S and M Vaysberg (2013). Antibodies to matrix metalloproteinase 9. Google Patents.
94. Bendell, JC, A Starodub, MA Shah, S Sharma, ZA Wainberg and DL Thai (eds.) (2015). Phase I study of GS-5745 alone and in combination with chemotherapy in patients with advanced solid tumors. *ASCO Annual Meeting Proceedings.*
95. Wu, AM and PD Senter (2005). Arming antibodies: prospects and challenges for immunoconjugates. *NatureB iotechnology*, 23(9), 1137–1146.
96. Foda, HD and S Zucker (2001). Matrix metalloproteinases in cancer invasion, metastasis and angiogenesis. *Drug Discovery Today*, 6(9), 478–482.
97. Andreini, C, L Banci, I Bertini, C Luchinat and A Rosato (2004). Bioinformatic comparison of structures and homology-models of matrix metalloproteinases. *Journal of Proteome Research*, 3(1), 21–31.
98. Remacle, AG, VS Golubkov, SA Shiryaev, R Dahl, JL Stebbins, AV Chernov, AV Cheltsov, M Pellecchia and AY Strongin (2012). Novel MT1-MMP small-molecule inhibitors based on insights into

hemopexin domain function in tumor growth. *Cancer Research*, 72(9), 2339–2349.
99. Zhu, L, P Kate and VP Torchilin (2012). Matrix metalloprotease 2-responsive multifunctional liposomal nanocarrier for enhanced tumor targeting. *ACS Nano*. 6(4), 3491–3498.
100. Zhang, J, Z-F Yuan, Y Wang, W-H Chen, G-F Luo, S-X Cheng, RX Zhuo and XZ Zhang (2013). Multifunctional envelope-type mesoporous silica nanoparticles for tumor-triggered targeting drug delivery. *Journal of the American Chemical Society*, 135(13), 5068–5073.

8. Matrix Metalloproteinases in Cardiovascular Diseases

Merry L. Lindsey* and Peter Libby†

**Professor of Physiology and Medicine, Mississippi Center for Heart Research, University of Mississippi Medical Center, Jackson, MS, USA*
†Division of Cardiovascular Medicine, Department of Medicine, Brigham and Women's Hospital, Harvard Medical School, Boston, MA, USA

8.1 Introduction

Remodeling of the extracellular matrix (ECM) has proven pivotal in the development and progression of many cardiovascular diseases (CVDs). Disordered metabolism of the ECM characterizes both diseases of the vasculature and of the myocardium. While many proteolytic enzymes may contribute to ECM degradation (e.g., cysteinyl, serine, and carboxyl proteinases), matrix metalloproteinases (MMPs) and their inhibitors have undergone the most intense scrutiny in this regard. Biological roles of the cardiovascular ECM coordinated by MMPs include integrating mechanical signals; mediating cell communication; promoting cell survival or stimulating apoptosis; providing structural integrity to blood vessels and the myocardium; regulating diastolic stiffness properties; remodeling in response to injury, inflammation, or growth; sequestering cytokines and growth factors; and serving as a scaffold to anchor cells (Table 8.1).[1] This

Table 8.1. Biological roles of the ECM in the cardiovascular system.

Integrate mechanical signals and transmit force
Mediate cell adhesion, cell–cell communication, and cell phenotype
Promote cell survival or stimulate apoptosis
Provide structural integrity to blood vessels and myocardium
Regulate diastolic stiffness properties of the ventricles
Remodel in response to injury, inflammation, or growth
Process cytokines and growth factors

Source: Modified from Liu *et al.* (2006).

chapter summarizes some of the vast literature regarding roles of MMPs and tissue inhibitors of matrix metalloproteinases (TIMPs) in atherosclerosis, arterial aneurysm formation, myocardial infarction (MI), and heart failure (HF). We aim to provide an illustration of the fundamental principles regarding the participation of MMPs in CVD rather than an encyclopedic review.

8.2 Atherosclerosis

Atherosclerosis, a common chronic inflammatory condition, involves extensive ECM remodeling during all of its phases.[2] Atherosclerotic plaques develop over many decades. While much concern has focused on the production of bulky lesions that protrude inward and impede blood flow, during much of the life history of an atheroma the plaque grows outward in an ablumenal direction rather than inward, a direction that can produce arterial stenosis. Thus, the early phases of atherogenesis involve outward remodeling. Arteries normally have a trilaminar structure consisting of an inner tunica intima, a middle medial layer, and an outermost layer, the adventitia. A monolayer of endothelial cells lines the intima and provides a contact surface with the blood.

In humans, the intima typically contains ECM with some resident arterial smooth muscle cells (SMCs) in the normal state. An internal elastic lamina comprised of mature, cross-linked elastin demarcates the intima from the tunica media. This middle layer of human arteries consists primarily of layers of SMCs sandwiched between ECM. Obstructive atherosclerosis usually affects muscular arteries, vessels which contain SMCs embedded in a matrix of collagen, elastin, proteoglycan, and intermediate filaments in less demarcated layers than in elastic arteries. In elastic arteries, the typical site of aneurysmal disease, more organized concentric bands of elastin separate layers of SMCs. The external elastic lamina forms the boundary between the media and the adventitia. The adventitia contains a loose ECM, mast cells, nerve endings, small vessels known as vasa vasorum that can penetrate into the outer third of the tunica media, and occasional lymphoid tissue.[3]

Outward expansion of growing lesions almost certainly requires dissolution of ECM and remodeling of the elastic layers. Vascular SMCs can produce a panel of MMPs capable of such ECM remodeling.[4] Their expression of MMPs responds to inflammatory mediators such as cytokines. Proteinases produced by SMCs plausibly produce some of the remodeling that permits outward expansion of nascent atheromata as they take root in the artery wall.[5]

Recent experiments in mice that develop atherosclerotic lesions due to mutations that render them hyperlipidemic have provided some mechanistic insight into the role of MMPs in outward remodeling of arteries during atherogenesis. Animals that lack the signaling receptor for interleukin-1 (IL-1) exhibit a failure to remodel, yielding lesions that encroach upon the arterial lumen at an earlier stage than mice wild type for the IL-1 receptor.[5] Array experiments in vascular SMCs pointed to MMP-3 as an exquisitely IL-1-responsive proteinase. Further *in vitro* and *in vivo* experiments implicated MMP-3 as a causal enzyme in expansive remodeling of the arterial ECM during plaque evolution. Either isoform of IL-1, alpha and beta, can induce MMP-3 production by vascular SMC.[6]

During atherogenesis and in response to arterial injury such as that produced by contemporary percutaneous therapeutic interventions, SMCs migrate from the media into the intima. This process requires traversing the internal elastic lamina and the other components of the arterial ECM. Again, MMPs produced by SMCs may facilitate this transit through the matrix barriers to allow SMCs to take up residence in the intima. Here, joined by resident intimal

SMCs, further matrix remodeling can lead to the formation of a fibrous cap comprised of SMCs and the collagenous ECM that they produce. This structure overlies the lipid core that forms in evolving atheromata. Experimental interventions that impair MMP activity can impede these remodeling processes. Administration of MMP inhibitors, which result in an increase ("gain of function") of tissue inhibitor of metalloproteinases (TIMPs) –1 or –2, and deficiencies in various MMPs have established a role for these enzymes in these aspects of arterial remodeling during experimental atherogenesis.[7–9] The effects of manipulation of MMP activity on the size of plaques have proven complicated.[10] Inactivation of TIMP-1 in atherosclerosis-prone mice yielded smaller plaques. The lesions in these mice accumulated more macrophages. Presumably MMP activity could impair accumulation of ECM in plaques, yielding a decrease in their volume. Mice with double deficiency of MMP-3 and apolipoprotein E (ApoE, hence atherosclerosis-prone), showed increased plaque volume and collagen content. Transgenic mice that have a "gain of function" of MMP-1 directed by a macrophage promoter also have smaller plaques and less collagen accumulation.[11] Thus, MMPs may have a net effect of reducing plaque volume while promoting arterial remodeling, both outward expansion and smooth muscle migration with fibrous cap formation. These effects can result from the abilities of MMPs to degrade the ECM as well as from their action on non-ECM substrates. As MMPs can either activate and/or inactivate proinflammatory cytokines and can also modulate integrin functions, the effects of manipulation of MMP gene expression during atherogenesis may involve more than ECM degradation, for example, modulating inflammatory responses or adhesive interactions implicated in atherogenesis.[12]

Regions of mature atherosclerotic plaques develop a rich microvasculature. Angiogenesis likely initiates the creation of these neovessels. This process requires microvascular endothelial cells to penetrate the dense layers of ECM in the medial and intimal layers of the artery.[13] Regional hypoxic stimuli drive this neovessel formation, which likely involves MMPs produced by the endothelial cells.

Hypoxic conditions, reactive oxygen species, and inflammatory mediators can stimulate the production of MMPs by ECs.[14]

The thrombotic complications of atherosclerosis cause the most dramatic clinical complications. One common form of plaque disruption involves a fissure or fracture of the plaque's fibrous cap. Interstitial collagen endows the plaque's fibrous cap with tensile strength that confers biomechanical stability. Considerable evidence suggests that MMPs mediate the degradation of interstitial collagen in the plaque, which can render the cap fragile and susceptible to rupture, thus provoking thrombosis.[8,15] Atheromatous plaques in humans overexpress many MMPs including the interstitial collagenases MMPs-1 and -13 (Fig. 8.1).[16] Such plaques contain partially degraded collagen indicative of the action of interstitial collagenases.[16] Macrophages appear to produce the bulk of the MMPs in these atheromatous, lipid-rich plaques.[16,17] The local regulation of MMP expression in atheroma depends on the shear stress experienced by the intimal endothelial layer. Plaques that form in regions of low shear stress overexpress MMPs in experiments performed in atherosclerotic pigs.[18,19]

An extensive series of experiments in mice have affirmed that MMPs can govern the collagen content of plaques (Table 8.2). Experiments in mice have shown that lack of function of MMP-8 or MMP-13 yields accumulation of collagen in atheromata.[20–22] Mice with a mutated collagenase-resistant form of interstitial collagen also have a reinforced collagenous structure.[23] Thus, both loss-of-function and gain-of-function manipulations in mice affirm MMPs as major regulators of plaque collagen content. *Ex vivo* studies of human atherosclerotic plaques also implicate MMP-13 as well as MMPs-1 and –8 in plaque collagenolysis.[21,22] Experiments in mice susceptible to atherosclerosis and doubly deficient in MMP-8 and MMP-13 suggest a dominant role for MMP-13 in this species (MMP-1 is the prominent interstitial collagenase in humans.)[22] Deficiency of MMP-14 in bone-marrow-derived cells yields increased plaque collagen in atherosclerotic mice, likely due to impaired activation of MMP-13.[24]

Lipid lowering, an intervention that can reduce the risk of thrombotic events in humans, may limit macrophage expression of

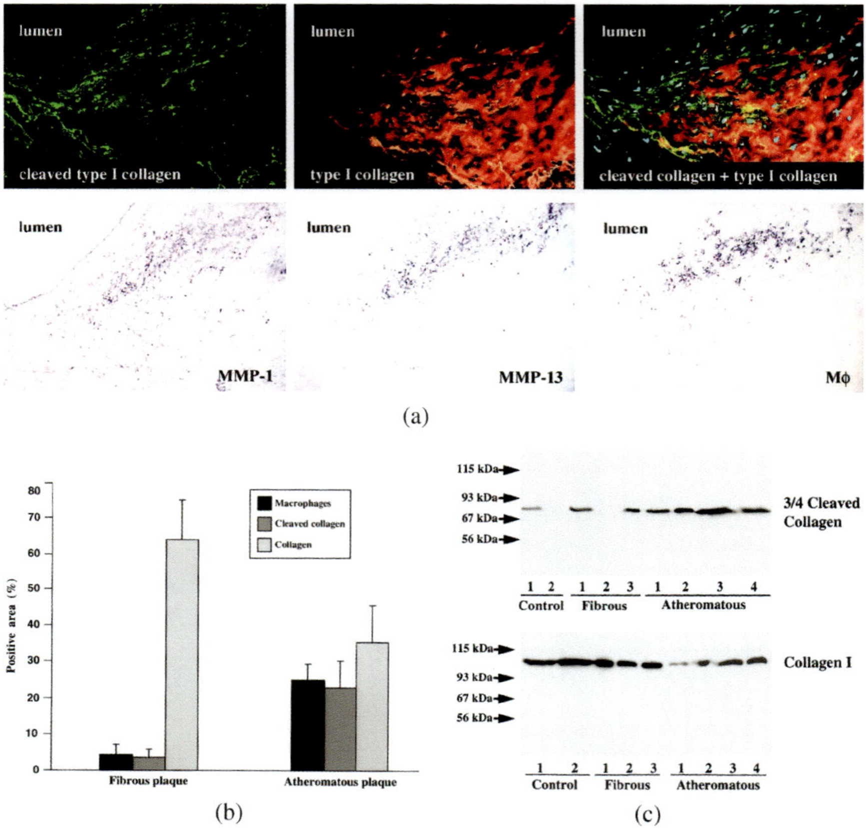

Figure 8.1. Elevated MMP collagenolytic capacity in human atherosclerotic plaques. (a) Immunofluorescence for collagenase-cleaved type I collagen neoepitope (green; top left) and collagen type I (Texas red; middle) provides *in situ* evidence for collegenolysis in the shoulder area of a typical human plaque obtained at carotid endarterectomy. Top right, double immunolabeling for both antigens (nuclei were counterstained with bisbenzimide, yielding blue fluorescence). Adjacent sections underwent immunohistochemical staining for the interstitial collagenases MMP-1 and -13 as well as for human macrophages (Mϕ) (bottom). (b) Shows quantitative color image analysis of the positively stained area for the indicated targets in a series of human atheromata classified a prior by morphologic criteria (mean ± SD, $n = 6$). (c) Extracts (50 µg of total protein/lane) from two nonatherosclerotic arteries (control) and three fibrous and four atheromatous plaques underwent western blotting to assess protein expression of collagenase-cleaved type I collagen as well as of collagen type I. (From Sukhova *et al.*, 1999.)

Table 8.2. MMPs regulate collagen content of mouse atheromata.

Intervention	Effect on Plaque Collagen Content	Reference
Collagenase resistance	Increases	23
MMP-13 inhibition	Increases	21
MMP-13 deficiency	Increases	20
MMP-8/13 double deficiency	Showed collagenolysis MMP-13 > MMP-8	22
MMP-14 bone marrow cell deficiency	Increases	24

MMPs in animals, and perhaps in human plaques as well.[25–27] Extreme dietary restriction of hyperlipidemia in rabbits can also limit MMP expression and yield collagen accumulation. These results provide mechanistic insight into lipid-lowering therapies that can mitigate the complications of atherosclerosis. Such data support the notion that MMPs participate causally in the thrombotic complications of atherosclerosis (Fig. 8.2).

In addition to fracture of the fibrous cap, a superficial erosion of the endothelial layer of the intima can also precipitate thrombotic complications of atherosclerosis.[28,29] The mechanisms of superficial erosion have received much less attention than plaque rupture, considered the dominant form of plaque disruption yielding thrombosis.[29,30] Superficial erosion accounts for a substantial minority of thrombotic complications of atherosclerosis, particularly in younger individuals and those with diabetes, a growing segment of the population with acute thrombotic complications of atherosclerosis.[31] Severing of the tethers of endothelial cells to the subjacent membrane mediated by MMPs might contribute to superficial erosion. Membrane type 1-MMP (MT1-MMP) (MMP-14), an activator of MMP-2, a type IV collagenase, can increase in endothelial cells exposed to oxidatively modified low-density lipoprotein (LDL).[32] As type IV collagen abounds in the subendothelial basement membrane, local activation of MMP-2 might promote sloughing or desquamation of endothelial cells or loosen their attachment to the subjacent

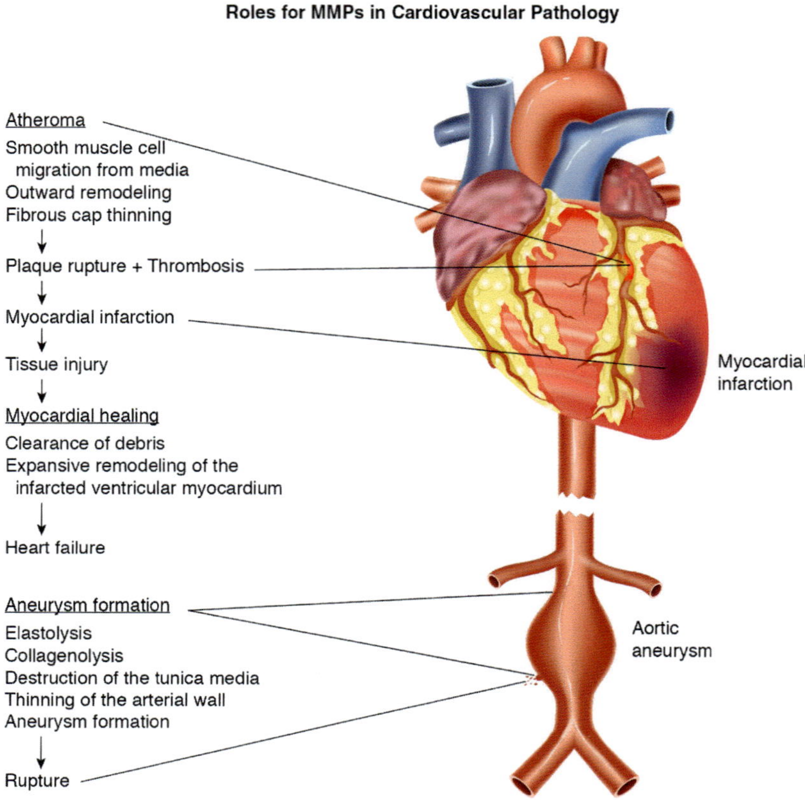

Figure 8.2. Roles for MMPs in cardiovascular pathology. Atherosclerosis causes the majority of HFrEF due to ischemic cardiomyopathy. During atheroma formation, proteinases prove pivotal in the migration of SMCs from the tunica media, the middle layer of arteries, into the usually thin intimal layer where they can proliferate, hypertrophy, elaborate ECM, and contribute to atheroma formation. As atherosclerotic plaques evolve, they grow outward for the first part of their life history. This process requires remodeling of the ECM mediated by proteinases including MMP-3 as revealed by experimental studies in mice (see text). Overlying the plaque's lipid-rich central core, a fibrous cap forms during atherogenesis. Excessive production of collagenolytic MMPs by plaque macrophages (lipid laden foam cells) can degrade the interstitial collagen that confers strength on the plaque's fibrous cap rendering the structure thin and susceptible to rupture. Disruption of thin-capped atherosclerotic plaques often triggers fatal thrombosis by allowing blood to contact highly thrombogenic material produced by macrophages in the plaque core. The thrombi, when sustained and occlusive, can lead to MI, a process that causes ischemic tissue injury. As depicted in Fig. 8.3, the elaboration of proteinases by proinflammatory

basement membrane sufficiently to provoke endothelial cell death by anoikis. Multiple genome-wide association studies have implicated variations in the collagen IV gene with atherosclerosis, another indication of the potential importance of the MMP-14/MMP-2 axis in CVD.[33] Together these concepts point to participation of MMPs in both major forms of disruption of atherosclerotic plaques that yield thrombosis: fibrous cap fracture and superficial erosion of the endothelial monolayer.

In summary, MMPs, by acting on matrix and nonmatrix substrates, participate in arterial ECM remodeling at all phases of atherosclerosis from inception of the lesions through evolution of the plaque and ultimately the thrombotic complications of these lesions. Lipid-lowering therapy and potentially direct anti-inflammatory interventions may mute MMP actions and mitigate some of the clinical complications of this common disease.[8]

8.3 MMPs in Arterial Aneurysm Formation

Considerable evidence implicates MMPs in the generation of arterial aneurysms.[34–37] While aneurysmal disease can affect all arteries, the largest literature implicating MMPs in aneurysm genesis focuses on

Figure 8.2. (*Continued*) leukocytes mediate much tissue damage, but also permit clearance of debris due to death of cells and destruction of the ECM. Expansive remodeling of the infarcted ventricular myocardium associates with poor prognosis and the development of HF. Weakening of the myocardial ECM due to proteolysis favors this geometrical remodeling, a common precursor to the development of chronic HF in ischemic cardiomyopathy. Remodeling of the ECM can also contribute to the pathogenesis of nonischemic forms of cardiomyopathy as well. Proteases of various classes including MMPs contribute to the formation of arterial aneurysms as exemplified in this diagram by an infrarenal AAA. Transmural destruction of the artery wall with fragmentation and ultimately disintegration of the elastic laminae set the stage for this collapse of the architecture of the arterial wall. As the radius of the artery increases, so does wall stress according to the Laplace principle. This positive feedback situation favors rupture of the aneurysm, too often a fatal event. Imbalance between MMPs and other proteolytic enzymes and their endogenous inhibitors undoubtedly contributes to the structural changes that favor aneurysm formation and complication.

a common form, abdominal aortic aneurysm (AAA). AAA commonly associates with atherosclerosis, tobacco abuse, and male sex. Genetic defects in ECM metabolism (e.g., fibrillin) more commonly cause aneurysmal disease of the thoracic aorta or cerebrovascular circulation. The hereditary syndromic arterial aneurysmal conditions thus seem to involve defects in MMP substrates more often than the enzymes themselves.[38] Therefore, this discussion will focus on atherosclerotic AAA.

The hallmark of AAA formation, transmural depletion and disorganization of elastin, implicates elastolysis by mere microscopic inspection of the lesions. Despite the seemingly evident implication of elastolytic enzymes in AAA formation, this hypothesis requires the careful consideration of a number of obstacles. Specimens of human AAA tissue obtained at surgery represent late stage disease, the "burned out embers" of the preceding pathological processes. While analyses of the "neck" regions that adjoin the aneurysmal tissue with the more normal portions of the aorta permit a glimpse of what might occur at earlier stages of aneurysm formation, we have a limited ability to analyze the formative stage of these arterial swellings.

While a number of experimental preparations exist that produce lesions with more or less characteristic of human AAA disease, each of the commonly used animal "models" has limitations, and they may not faithfully replicate the processes at work in the pathogenesis of human AAA.[39] Nonetheless, a large and consistent clinical and experimental literature implicates MMPs and other matrix-degrading enzymes in the pathogenesis of aneurysms (Fig. 8.2). Analysis of human lesions shows overexpression of virtually all MMPs studied, in addition to elastolytic cathepsins and serine proteinases such as neutrophil elastase (Table 8.2). While inconsistent, some studies have shown reciprocal decreases in the expression of TIMPs in aneurysmal lesions.[35] In the experimental arena, both gain-of-function and loss-of-function experiments in various animal preparations support a pathogenic role of MMPs in aneurysm formation.[40–45]

Human AAA display evidence of an immune component mediated predominantly by T helper cell 2 (Th2) adaptive immune response.[46] Experimental studies showed that absence of the signature Th1 cytokine, interferon gamma, augments aneurysm formation, in association with Th2 polarization. The Th2 cytokine IL-4 boosts MMP-9 and MMP-12 production in arterial tissue, providing a link with elastolysis mediated by MMPs and the pathogenesis of aneurysmal disease.[44,45,47] Nonmetalloproteinases also likely work in concert with MMPs to promote ECM degradation in arterial diseases including aneurysms. In particular, a large body of work implicates the cysteinyl elastases (cathepsins S, K, and L) in aneurysm formation.[48,49] Serine proteinases may also participate in ECM remodeling in the context of aneurysmal disease.[50] In summary, largely consistent human and experimental literature supports excessive activity of MMPs as pathogenic in the genesis of aneurysms.

8.4 Myocardial Infarction

MI occurs after the occlusion of an atherosclerotic coronary artery, resulting in necrosis of the cardiomyocytes downstream of the occlusion site.[51] MI involves ECM remodeling in both the infarct and remote noninfarcted regions throughout the time continuum of the inflammatory and scar-forming processes during the wound-healing response. Remodeling of the left ventricle (LV) following MI involves ECM turnover and augmented expression of several MMPs and TIMPs (Figs. 8.2 and 8.3). Collagen dissolution post-MI facilitates the removal of necrotic cardiomyocytes in the first days following coronary artery occlusion. Within the first week of an evolving infarct, laying down of a new ECM begins to counterbalance this heightened proteolysis, and generates granulation tissue and ultimately a scar that can enhance structural stability.[1,52] If ECM degradation exceeds its synthesis, adverse expansive infarct remodeling, LV aneurysms, or, in the extreme, rupture can occur.[53] Following permanent coronary artery occlusion in mice, rupture accounts for one-third to one-half of post-MI deaths and peaks about days 3–7 post-MI, a time of prime structural vulnerability for the evolving infarct. In contrast, if ECM synthesis exceeds degradation or continues past the

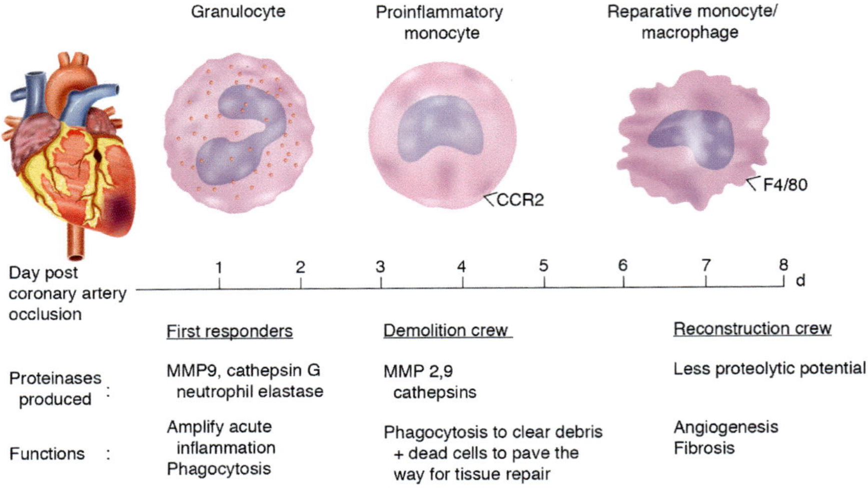

Figure 8.3. Distinct profiles of proteolytic activity critically influence the functional capacities of leukocyte subclasses during myocardial infarction. This timeline depicts roughly the participation of various classes of leukocytes in myocardial ischemic injury and its repair. In the first day or two following coronary artery occlusion, granulocytes predominate in the leukocyte population in the evolving infarct. These first responders express MMP-9, cathepsin G, proteinase 3, and neutrophil elastase. These proteinases can contribute to tissue injury. The granulocytes can also engage in an early phase of phagocytosis to clear the debris elaborated by dead and dying cells. In the following days, a population of proinflammatory monocytes predominates, cells characterized in the mouse by high levels of expression of the surface structure Ly6c and by the expression of chemokine receptor 2 (CCR2). These proinflammatory monocytes elaborate MMP-2 and MMP-9 as well as a variety of sulfhydryl proteinases of the cathepsin category that can amplify tissue damage and attack the ECM, paving the way for subsequent tissue repair. In the latter phase of acute ischemic injury, monocyte macrophages that display a reparative or proresolving panel of functions predominate. These leukocytes, characterized by lower level of expression of Ly6c in mice, bear the marker F4/80. They have less proteolytic potential than their proinflammatory counterparts, but elaborate mediators of angiogenesis such as vascular endothelial growth factor as well as stimuli to ECM protein production that favor fibrosis such as members of the TGF-β superfamily. Experimental studies in mice have shown that manipulation of these various functions exerted by leukocyte subclasses myocardial healing, geometrical remodeling, and ultimately cardiac function. Proteinases produced by these leukocytes play pivotal roles in these processes.

point of stabilization, an overly fibrotic myocardium can increase LV stiffness and provide an arrhythmogenic substrate.[54] In addition to collagen, other MI-relevant ECM substrates of MMPs include fibronectin, laminins, and osteopontin, which can regulate both ECM and inflammatory components of LV remodeling.[55–58]

To date, literature confirms increased expression during MI of MMP-1, -2, -3, -7, -8, -9, 12, 13, and 14 as well as TIMP-1, -2, and -3 (Tables 8.2 and 8.3).[53] Some MMPs and TIMPs demonstrate a shift in cell source post-MI. For example, MMP-28 from cardiomyocytes decreases post-MI, while MMP-28 from macrophages increases.[59] Of note, over a dozen MMPs remain unevaluated in the post-MI LV.

MMPs and TIMPs increase immediately after MI, due to both increased synthesis by endogenous cells and by increased influx of leukocytes as part of the inflammatory response. Neutrophils, first-responder leukocytes, contribute MMP-8, MMP-9, and MMP-12 as well as TIMP-1 (Fig. 8.3).[60,61] Macrophages provide a rich source of MMP-1 (in humans), MMP-3, MMP-7, MMP-8, MMP-9, MMP-12, MMP-14, and MMP-28 as well as all four TIMPs.[59,62,63]

Cardiac myocytes express MMP-1 (in humans), MMP-2, MMP-3, MMP-7, MMP-9, MMP-14, TIMP-1, and TIMP-4.[53] The high expression of TIMP-4 in cardiomyocytes leads to a dramatic decline of total TIMP-4 post-MI in the infarct region as a result of myocyte loss. Other endogenous cells in the myocardium include endothelial cells and vascular SMCs, both of which express both MMP-2 and MMP-9.[53]

Cardiac myofibroblasts express MMP-1 (in humans), MMP-2, MMP-9, MMP-13, and MMP-14, although at different concentrations than leukocytes.[64] For example, leukocytes predominantly produce MMP-9, while MMP-2 (both intracellular and extracellular forms) remains elevated post-MI primarily due to increased expression by cardiomyocytes and cardiac fibroblasts.[65–68] Cardiac myocytes and fibroblasts constitutively produce pro-MMP-2. A multitude of stimuli augment the expression of this enzyme, including interleukin-1 beta (IL-1β), tumor necrosis factor-alpha (TNF-α), transforming growth factor-beta (TGF-β), mechanical load, and oxidative stress.[64,69] While macrophages express MMP-3, this proteinase also remains elevated post-MI through increased expression by cardiac fibroblasts.[70] In addition to blocking MMP activity, TIMPs stimulate fibroblast proliferation and phenotypic differentiation of fibrocytes into myofibroblasts.[71] TIMP-2 associates with increased fibroblast collagen synthesis, while both TIMP-3 and TIMP-4 inhibit cell growth and trigger myofibroblast apoptosis.[71]

The local inflammatory reaction remains a considerable contributor to MMP upregulation in the post-MI infarct region.[1] For example, excess TNF-α concentrations in the LV directly relate to increased local production of MMP-2 and MMP-9.[72] An extensive feedback loop also operates among MMPs and cytokines. For example, IL-1β stimulates MMP production in macrophages and fibroblasts, and MMP-2, MMP-3, and MMP-9 can proteolytically process IL-1β to its active form.[12] Prolonged incubation of the mature form of IL-1β with MMP-3 results in IL-1β degradation, demonstrating the ability to both turn on and turn off cytokine activity. Chemokines, another class of inflammatory molecules, remain regulated by MMPs and regulate MMPs.[73,74] By day 28 post-MI in the mice, inflammation and MMP expression showed low levels in the infarct region, signaling resolution.[75]

MMP-9 appears to have a major role in the pathogenesis of MI (Fig. 8.3). MMP-9 remains elevated post-MI, and previous studies show elevated levels in experiments with mice, rats, pigs, rabbits, and dogs, as well as humans.[76] The elevation in MMP-9 tracks with neutrophil influx.[77] Further, deletion of MMP-9 reduces macrophage infiltration and attenuates collagen accumulation and LV dilation post-MI.[76] In mice, MMP-9 deletion stimulates neovascularization post-MI and improves outcomes.[78] In contrast, the transgenic overexpression of MMP-9 only in macrophages also improves LV remodeling post-MI, thereby indicating dual roles for MMP-9, depending on when it is expressed and whether it originates from the neutrophil or the macrophage.[79]

TIMP-3 deficient mice show significantly elevated mortality post-MI due to enhanced ECM degradation and inflammatory cytokine expression that accelerates the development of systolic dysfunction.[80] Injection of vascular SMCs transfected with TIMP-3 into the peri-infarct region at 3 or 14 days post-MI reduced MMP-2 and MMP-9 activities, infarct wall thinning, and the development of LV dysfunction.[81]

In summary, extensive evidence shows involvement of about half of the MMP family members and all of the TIMPs in the ischemically injured myocardium. Changes in the balance between levels of MMPs and their inhibitors likely influence critical aspects of infarct healing that determine the clinical course both acutely and in the long term.

8.5 Heart Failure

HF, the inability of the heart to adequately pump and supply the rest of the body with sufficient oxygen, remains a common downstream consequence of a myriad of CVDs, including hypertension, MI, infiltrative processes, and various cardiomyopathies. HF has a strong inflammatory component and associates with the increased activities of MMPs and TIMPs. MMPs and TIMPs remain elevated in humans and experimental animals with HF, including MMP-2, MMP-3, MMP-8, MMP-9, MMP-19, and all four TIMPs.[82,83] While previous data show elevated levels of TIMP-1 in HF, other reports indicate plasma values remain lower in patients with HF. Sampling variabilities likely account for part of this apparent discrepancy. The balance between proteinase and antiproteinase activity remains in flux over time during the course of HF progression.

MI remains a common cause of HF (Fig. 8.3). Indeed, 70% of HF with reduced ejection fraction attributes MI as the underlying etiology, a condition referred to as ischemic cardiomyopathy.[84–91] Following MI, MMPs increase (as discussed previously) during the initial response, followed by a second wave during the progression to HF. This biphasic augmentation of MMPs and TIMPs requires consideration when designing experiments in animals and exploring therapeutic targets in clinical trials. In a cross-sectional analysis of 29 subjects with heart failure with reduced ejection fraction (HFrEF) and 29 subjects with heart failure with preserved ejection fraction (HFpEF), ECM and inflammatory proteins, including MMP-2 and MMP-9, showed similar concentrations.[92] While these MMPs localized in similar concentrations in these two types of HF, an apparent heterogeneity exists in the mechanisms that mediate either the release or clearance of these MMPs.

In summary, HF associates with augmented activity of MMPs, in part due to the inflammatory aspects of this disease. At this point, measurements of MMPs and their inhibitors in humans with HF serve primarily to provide insight into the pathophysiology of different forms of HF and other CVDs, as discussed in the following. Yet, in addition to potential utility as diagnostic and prognostic indicators, MMPs and TIMPs may also indicate response to therapy.[93]

8.6 MMPs and TIMPs as Prognostic Indicators of CVD

Previous studies have evaluated MMPs and TIMPs as plasma markers of cardiovascular status both in community-based populations and in CVD risk patients. Sundstrom and colleagues performed the first community-based investigation, using the Framingham Heart Study cohort to relate plasma MMP-9 and TIMP-1 to CVD risk factors and to LV structure and function by echocardiographic variables.[94–97] Their data showed that plasma MMP-9 associated with increased LV diastolic dimensions and wall thickness, which support the use of MMP-9 as a plasma marker of cardiac ECM degradation. MMP-9 correlated with smoking and diabetes, indicating likely cardiac and extra-cardiac sources. Plasma TIMP-1 also directly associated with the Framingham Risk Score and inversely correlated to LV systolic function. In addition to smoking and diabetes, plasma TIMP-1 correlated with age, body mass index, and the total over high-density lipoprotein (HDL) cholesterol ratio. In the Framingham cohort, higher circulating TIMP-1 associated with mortality risk, showing a multivariable-adjusted hazards ratio of 1.72 (95% CI, 1.30–2.27).[98]

In a cross-sectional evaluation of the Prospective Investigation of the Vasculature in Uppsala Seniors (PIVUS) community-based cohort of 70-year-old men and women with no valvular disease, MI, or HF (n = 891), serum MMP-9 weakly associated with LV wall thickness and isovolumic relaxation time.[99] Serum TIMP-1 positively correlated with LV mass and wall thickness and negatively correlated with ejection fraction, indicating that TIMP-1 may regulate LV systolic function. In the prospective Uppsala Longitudinal Study of Adult Men (ULSAM) cohort, serum MMP-9 and TIMP-1 both associated with risk of all-cause mortality.[100] TIMP-1 correlated with risk of CV mortality and stroke. Although adjustment for CVD risk factors attenuated these relationships, the associations affirm the implication of strong relationships among MMPs and TIMPs and the pathogenesis of CVD. In fact, multiple CV risk factors associate with increased MMP activity, including smoking, diabetes, homocysteine, and lipids.[1] Each of these factors stimulates cytokine release from activated leukocytes, which in turn stimulates the synthesis and activation of MMPs. MMPs can then proteolytically process cytokines and chemokines to either increase or diminish their activity, furnishing possible positive or negative feedback pathways.

One study in patients with existing coronary artery disease highlights the role of MMP-9. It showed that plasma MMP-9 rivaled the established biomarkers troponin and C reactive protein as a prognostic indicator.[101] The lowest quartile of MMP-9 associated with $<2.5\%$ event rate, while the highest quartile had $>15\%$ event rate. In 1979 patients with coronary artery disease, those who subsequently experienced a fatal CV event had higher mean concentrations of plasma TIMP-1, with an age and sex adjusted hazard ratio of future CV death being 1.37 (95% CI: 1.17–1.61, $p < .001$).[102] In summary, MMPs and TIMPs, both singly or as part of a panel of plasma markers, show promise as diagnostic and prognostic indicators of CVD and affirm the participation of these proteinases in the pathogenesis of these diseases.

8.7 Direct and Indirect MMP Inhibition

The known roles of numerous MMPs and TIMPs in atherosclerosis, aneurysms, MI, and HF (Table 8.3) provide a strong rationale for considering MMP inhibitors for the therapy of certain CVDs.

Table 8.3. Matrix metalloproteinases and tissue inhibitors of metalloproteinases with known associations to cardiovascular diseases.

	Atherosclerosis	Aneurysm	Myocardial infarction	Heart failure
MMP-1	+		*	+
MMP-2	+	+	+	+
MMP-3	+	+	+	+
MMP-7	+		+	
MMP-8	+	+	+	+
MMP-9	+	+	+	+
MMP-10	+			
MMP-12	+	+	+	
MMP-13	+		+	
MMP-14	+		+	
MMP-19			+	+
MMP-28			*	
TIMP-1	+		+	*
TIMP-2	+		+	+
TIMP-3	+		+	+
TIMP-4			–	+

+, increases; –, decreases; *, increases or decreases, depending on time of evaluation or cell type evaluated.
Note that information is not currently available for many of the MMPs and TIMPs.
Source: [26,51–55,10,24,35,114–120]

Specific MMP roles in CVD have undergone evaluation *in vivo* in mice with genetic deletion or transgenic overexpression of MMP-2, MMP-3, MMP-7, MMP-8, MMP-9, MMP-14, MMP-28, and TIMP-1, TIMP-2, TIMP-3, and TIMP-4, supporting the development of selective MMP antagonists or in some cases, agonists.[53,76,103–107] A variety of approaches permit pharmacological MMP inhibition: broad-spectrum MMP inhibitors, relatively specific MMP inhibitors, and specific inhibitors developed for MMP-9 and MMP-12.[60,108–113]

Despite setbacks, a lasting conceptual appeal and rationale have boosted the notion that pharmacologic inhibition of MMPs might limit atherosclerosis progression, aneurysm growth, or LV remodeling post-MI leading the HF. In humans, for example, increased dimensions of aortic aneurysms associate with increased risk of rupture and increased cardiac dimensions post-MI associate with increased risk of HF development. Rupture of aortic aneurysms and HF both have high case-fatality rates, hence the rationale of preventing aneurysm expansion or HF pharmacologically. Based on these findings, some MMPs (e.g., MMP-12) may warrant either an antagonist or agonist, depending on both the CVD involved (atherosclerosis vs. MI) and the time of administration of the therapy. Johnson and colleagues showed a protective effect of an MMP-12 inhibitor in preventing atherosclerosis development in ApoE-null mice, whereas Iyer *et al.* showed that early MMP-12 inhibition (using the same inhibitor) exacerbated post-MI LV remodeling by delaying the resolution of the inflammatory response.[60,110] In both cases, macrophage function remained impaired; this effect proved beneficial in the setting of atherosclerosis and detrimental in the setting of MI. Understanding MMP roles in the various CVD processes, therefore, reveals how effects on the same cell type can impart opposite consequences at the tissue level, and underscores the importance of staging disease and timing of therapeutic intervention against MMPs.

In humans, the initial pharmacological MMP inhibitor trials had over-simplified designs, given the diversity of both MMPs and their functions.[121,122] At the time of the development of the first inhibitors, the full family of enzymes and similar families (e.g., a disintegrin and metalloproteinase [ADAM]) remained unknown.[123] The Prevention of Myocardial Infarction Early Remodeling (PREMIER) trial evaluated the efficacy of the MMP inhibitor PG116800 on post-MI remodeling.[124] This trial enrolled 253 patients who received treatment within 48 hours of presentation. While the initial design involved the treatment with 200-mg oral twice daily for 180 days, concerns for potential off-target musculoskeletal effects stimulated a switch to once-a-day treatment. The primary response variable was changes in LV end diastolic volumes (LVEDVs) from baseline to

90 days post-MI. In the placebo group, the change in LVEDV was 10.3%, while in the PG116800 group the change was 8.4% (p = .31). Of note, similar trials showed a much larger change in LVEDV, indicating a lack of dilation within the subjects in both groups of this trial.[121] This issue, along with the inadequate dosing regimen such that a large number of patients failed to achieve therapeutic inhibition, likely contributed significant confounders, thus making it difficult to rule in or out MMP inhibition as a post-MI therapy.

Doxycycline, a broad-spectrum antibiotic in the tetracycline class, can exert antiprotease activity in addition to antimicrobial actions.[125–127] The Food and Drug Administration approved doxycycline as an MMP inhibitor and studies have proven its effectiveness in preventing tissue destruction in the setting of gingivitis.[128,129] Observations in animals have supported the efficacy of doxycycline, in a number of experimental cardiovascular conditions ranging from aortic remodeling in experimental aneurysm expansion to post-MI remodeling.[130–132] In 60 patients scheduled for elective aneurysmal repair, two weeks of treatment with doxycycline suppressed aortic wall inflammation, resulting in 72% reduction in neutrophils and 95% reduction in cytotoxic T-cells in the aneurysm.[133] Pilot studies in humans have also provided conceptual support for this intervention. Yet, to date, large-scale clinical trials have not definitively established MMP inhibitors as effective therapies in this condition. The early short-term doxycycline therapy in patients with acute myocardial infarction and left ventricular dysfunction to prevent the ominous progression to adverse remodeling trial (TIPTOP) evaluated efficacy of treating post-MI patients, starting immediately after percutaneous intervention, with 100-mg doxycycline twice a day for 7 days.[134,135] Enrollment included a total of 110 patients, and the treated group showed decreased LVEDV compared to control (standard of care over the six months of follow-up).

In addition to direct MMP inhibition, CVD medications can also activate or inhibit MMPs.[105] Nitroglycerin increases expression and activity of MMP-2, MMP-7, and MMP-9, and decreases TIMP-1 in human macrophages.[136] Tissue plasminogen activators increase MMP-1, MMP-2, MMP-3, MMP-9, MMP-12, and MMP-14 as well as TIMP-1 and TIMP-2.[137,138] Likewise, anticoagulants increase MMP-2, MMP-9, and TIMP-2 concentrations and decrease TIMP-1 in the blood.[139,140]

Angiotensin-converting enzyme (ACE) inhibitors lower MMP-1, MMP-2, MMP-3, and MMP-9 and raise TIMP-1.[141,142] In addition to effects on MMP synthesis, the Yamamoto laboratory has shown that the ACE inhibitors imidapril, lisinopril, and captopril can bind directly to the active site of MMP-9 to inhibit enzymatic activity.[143–146] Further, imidapril showed strong inhibitory activity against MMP-9 in hamsters with MI, indicating that direct MMP inhibition may contribute to some of the clinical benefits of ACE inhibition.[143,147] Angiotensin II receptor inhibitors lower MMP-2, MMP-3, and MMP-9 while raising TIMP-1; aldosterone antagonists lower MMP-1, MMP-2, and MMP-9 and raise TIMP-2; and beta adrenergic receptor blockers lower MMP-2 and MMP-9 and raise TIMP-1, TIMP-2, and TIMP-3.[148]

Several statins inhibit MMPs, either directly or indirectly. This class of drugs indirectly lowers MMP activity by blunting inflammation.[149–152] Statins can also directly reduce MMP-1, MMP-2, MMP-9, and MMP-12 and increase TIMP-1 concentrations.[149–151] Simvastatin decreases plasma MMP-9 in hypercholesterolemic patients with coronary artery disease. Moreover, nonrandomized pravastatin treatment showed decreased MMP-2 and increased TIMP-1 in human carotid endarterectomy specimens compared to those not taking a statin. These alterations could render plaques less likely to rupture and provoke clinical events.[27,153]

Other CVD-related medications indirectly regulate MMP and TIMP activities. Nonsteroidal anti-inflammatory drugs lower MMP-1 and raise TIMP-1 concentrations.[154,155] Rosiglitazone decreases serum MMP-9 in patients with type 2 diabetes.[156] The calcium channel blockers amlodipine and diltiazem increase MMP-1, MMP-2, and TIMP-1 in cultured human umbilical vein endothelial cells.[157]

In summary, these data support the continued consideration of modulating MMPs in the therapy of both vascular and myocardial diseases. Nonetheless, challenges remain in identifying the particular proteinases involved in specific stages of the disease in question, and deploying novel selective inhibitors that avoid the undesired actions that have challenged the clinical application of broad-spectrum MMP inhibitors.

References

1. Liu, P, M Sun and S Sader (2006). Matrix metalloproteinases in cardiovascular disease. *Canadian Journal of Cardiology*, 22(Suppl. B), 25B–30B.
2. Libby, P, RE Gerszten and PM Ridker (2014). Biomarkers, proteomics, metabolomics, and personalized medicine. In *Heart Disease*, 10th Ed., D Mann *et al.* (eds.). Elsevier: Philadelphia.
3. Mohanta, SK, C Yin, L Peng, P Srikakulapu, V Bontha, D Hu, F Weih, C Weber, N Gerdes and AJ Habenicht (2014). Artery tertiary lymphoid organs contribute to innate and adaptive immune responses in advanced mouse atherosclerosis. *Circulation Research*, 114, 1772–1787.
4. Galis, Z, M Muszynski, G Sukhova, E Simon-Morrisey, E Unemori, M Lark, E Amento and P Libby (1994). Cytokine-stimulated human vascular smooth muscle cells synthesize a complement of enzymes required for extracellular matrix digestion. *Circulation Research*, 75, 181–189.
5. Alexander, MR, CW Moehle, JL Johnson, Z Yang, JK Lee, CL Jackson and GK Owens (2012). Genetic inactivation of IL-1 signaling enhances atherosclerotic plaque instability and reduces outward vessel remodeling in advanced atherosclerosis in mice. *Journal of Clinical Investigation*, 122, 70–79.
6. Beltrami-Moreira, M, A Vromman, GK Sukhova, EJ Folco and P Libby (2016). Redundancy of IL-1 isoform signaling and its implications for arterial remodeling. *PLoS ONE*, 11, e0152474.
7. Yla-Herttuala, S, JF Bentzon, M Daemen, E Falk, HM Garcia-Garcia, J Herrmann, I Hoefer, S Jauhiainen, JW Jukema, R Krams, BR Kwak, N Marx, M Naruszewicz, A Newby, G Pasterkamp, PW Serruys, J Waltenberger, C Weber, L Tokgozoglu, ESC Working Group of Atherosclerosis and Vascular Biology (2013). Stabilization of atherosclerotic plaques: an update. *European Heart Journal*, 34, 3251–3258.
8. Libby, P (2013a). Mechanisms of acute coronary syndromes and their implications for therapy. *New England Journal of Medicine*, 368, 2004–2013.

9. Newby, AC (2014). Proteinases and plaque rupture: unblocking the road to translation. *Current Opinion in Lipidology*, 25, 358–366.
10. Dollery, CM and P Libby (2006). Atherosclerosis and proteinase activation. *Cardiovascular Research*, 69, 625–635.
11. Lemaitre, V, TK O'Byrne, AC Borczuk, Y Okada, AR Tall and J D'Armiento (2001). ApoE knockout mice expressing human matrix metalloproteinase-1 in macrophages have less advanced atherosclerosis. *Journal of Clinical Investigation*, 107, 1227–1234.
12. Schonbeck, U, F Mach and P Libby (1998). Generation of biologically active IL-1 beta by matrix metalloproteinases: a novel caspase-1-independent pathway of IL-1 beta processing. *Journal of Immunology*, 161, 3340–3346.
13. Libby, P and U Schonbeck (2001). Drilling for oxygen: angiogenesis involves proteolysis of the extracellular matrix. *Circulation Research*, 89, 195–197.
14. Rajagopalan, S, XP Meng, S Ramasamy, DG Harrison and ZS Galis (1996). Reactive oxygen species produced by macrophage-derived foam cells regulate the activity of vascular matrix metalloproteinases in vitro. Implications for atherosclerotic plaque stability. *Journal of Clinical Investigation*, 98, 2572–2579.
15. Libby, P (2013b). Collagenases and cracks in the plaque. *Journal of Clinical Investigation*, 123, 3201–3203.
16. Sukhova, GK, U Schonbeck, E Rabkin, FJ Schoen, AR Poole, RC Billinghurst and P Libby (1999). Evidence for increased collagenolysis by interstitial collagenases-1 and -3 in vulnerable human atheromatous plaques. *Circulation*, 99, 2503–2509.
17. Galis, Z, G Sukhova, R Kranzhöfer, S Clark and P Libby (1995). Macrophage foam cells from experimental atheroma constitutively produce matrix-degrading proteinases. *Proceedings of the National Acadamy of Sciences USA*, 92, 402–406.
18. Chatzizisis, YS, AB Baker, GK Sukhova, KC Koskinas, R Beigel, M Jonas, AU Coskun, BV Stone, C Maynard, GP Shi, P Libby, CL Feldman, ER Edelman and PH Stone (2011). Augmented expression and activity of extracellular matrix-degrading enzymes in regions of low endothelial shear stress co-localize with coronary atheromata with thin fibrous caps in pigs. *Circulation*, 123, 621–630.
19. Koskinas, KC, GK Sukhova, AB Baker, MI Papafaklis, YS Chatzizisis, AU Coskun, T Quillard, M Jonas, C Maynard, AP Antoniadis, GP Shi, P Libby, ER Edelman, CL Feldman and PH Stone (2013).

Thin-capped atheromata with reduced collagen content in pigs develop in coronary arterial regions exposed to persistently low endothelial shear stress. *Arteriosclerosis and Thrombosis: A Journal of Vascular Biology*, 33, 1494–1504.
20. Deguchi, JO, E Aikawa, P Libby, JR Vachon, M Inada, SM Krane, P Whittaker and M Aikawa (2005). Matrix metalloproteinase-13/collagenase-3 deletion promotes collagen accumulation and organization in mouse atherosclerotic plaques. *Circulation*, 112, 2708–2715.
21. Quillard, T, Y Tesmenitsky, K Croce, R Travers, E Shvartz, KC Koskinas, G Sukhova, E Aikawa, M Aikawa and P Libby (2011). Selective inhibition of matrix metalloproteinase 13 (MMP-13) increases collagen content of established mouse atheromata. *Arteriosclerosis and Thrombosis: A Journal of Vascular Biology*, 31, 2464–2472.
22. Quillard, T, HA Araujo, G Franck, Y Tesmenitsky and P Libby (2014). Matrix metalloproteinase-13 predominates over matrix metalloproteinase-8 as the functional interstitial collagenase in mouse atheromata. *Arteriosclerosis and Thrombosis: A Journal of Vascular Biology*, 34, 1179–1186.
23. Fukumoto, Y, JO Deguchi, P Libby, E Rabkin-Aikawa, Y Sakata, MT Chin, CC Hill, PR Lawler, N Varo, FJ Schoen, SM Krane and M Aikawa (2004). Genetically determined resistance to collagenase action augments interstitial collagen accumulation in atherosclerotic plaques. *Circulation*, 110, 1953–1959.
24. Schneider, F, GK Sukhova, J Aikawa, J Canner, N Gerdes, S-MT Tang, G-P Shi, SS Apte and P Libby (2008). Matrix metalloproteinase-14 deficiency in bone marrow-derived cells promotes collagen accumulation in mouse atherosclerotic plaques. *Circulation*, 117, 931–939.
25. Aikawa, M, E Rabkin, Y Okada, SJ Voglic, SK Clinton, CE Brinckerhoff, GK Sukhova and P Libby (1998). Lipid lowering by diet reduces matrix metalloproteinase activity and increases collagen content of rabbit atheroma: a potential mechanism of lesion stabilization. *Circulation*, 97, 2433–2444.
26. Aikawa, M, E Rabkin, S Sugiyama, SJ Voglic, Y Fukumoto, Y Furukawa, M Shiomi, FJ Schoen and P Libby (2001). An HMG-CoA reductase inhibitor, cerivastatin, suppresses growth of macrophages expressing matrix metalloproteinases and tissue factor in vivo and in vitro. *Circulation*, 103, 276–283.

27. Crisby, M, G Nordin-Fredriksson, PK Shah, J Yano, J Zhu and J Nilsson (2001). Pravastatin treatment increases collagen content and decreases lipid content, inflammation, metalloproteinases, and cell death in human carotid plaques: implications for plaque stabilization. *Circulation,* 103, 926–33.
28. Bentzon, JF, F Otsuka, R Virmani and E Falk (2014). Mechanisms of plaque formation and rupture. *Circulation Research*, 114, 1852–1866.
29. Libby, P and G Pasterkamp (2015). Requiem for the 'vulnerable plaque'. *European Heart Journal*, 36(43), 2984–2987.
30. Libby, P, P Ganz, FJ Schoen and RT Lee (2000). The vascular biology of the acute coronary syndromes. In *Acute Coronary Syndromes*, 2nd Ed., EJ Topol (ed.), pp. 33–57. New York: Marcel Dekker.
31. Arbustini, E, B Dal Bello, P Morbini, AP Burke, M Bocciarelli, G Specchia and R Virmani (1999). Plaque erosion is a major substrate for coronary thrombosis in acute myocardial infarction. *Heart*, 82, 269–272.
32. Rajavashisth, TB, JK Liao, ZS Galis, S Tripathi, U Laufs, J Tripathi, NN Chai, XP Xu, S Jovinge, PK Shah and P Libby (1999). Inflammatory cytokines and oxidized low density lipoproteins increase endothelial cell expression of membrane type 1-matrix metalloproteinase. *Journal of Biological Chemistry*, 274, 11924–11929.
33. Turner, AW, M Nikpay, A Silva, P Lau, A Martinuk, TA Linseman, S Soubeyrand and R McPherson (2015). Functional interaction between COL4A1/COL4A2 and SMAD3 risk loci for coronary artery disease. *Atherosclerosis*, 242, 543–552.
34. Tamarina, NA, WD McMillan, VP Shively and WH Pearce (1997). Expression of matrix metalloproteinases and their inhibitors in aneurysms and normal aorta. *Surgery*, 122, 264–271; discussion 271–272.
35. Knox, JB, GK Sukhova, AD Whittemore and P Libby (1997). Evidence for altered balance between matrix metalloproteinases and their inhibitors in human aortic diseases. *Circulation*, 94, 205–212.
36. Wassef, M, BT Baxter, RL Chisholm, RL Dalman, MF Fillinger, J Heinecke, JD Humphrey, H Kuivaniemi, WC Parks, WH Pearce, CD Platsoucas, GK Sukhova, RW Thompson, MD Tilson and CK Zarins (2001). Pathogenesis of abdominal aortic aneurysms: a multidisciplinary research program supported by the National Heart, Lung, and Blood Institute. *Journal of Vascular Surgery*, 34, 730–738.

37. Thompson, RW and BT Baxter (1999). MMP inhibition in abdominal aortic aneurysms. Rationale for a prospective randomized clinical trial. *Annals of the New York Academy of Sciences*, 878, 159–178.
38. Loeys, BL, U Schwarze, T Holm, BL Callewaert, GH Thomas, H Pannu, JF De Backer, GL Oswald, S Symoens, S Manouvrier, AE Roberts, F Faravelli, MA Greco, RE Pyeritz, DM Milewicz, PJ Coucke, DE Cameron, AC Braverman, PH Byers, AM De Paepe and HC Dietz (2006). Aneurysm syndromes caused by mutations in the TGF-beta receptor. *New England Journal of Medicine*, 355, 788–798.
39. Libby, P (2015). Murine "model" monotheism: an iconoclast at the altar of mouse. *Circulation Research*, 117, 921–925.
40. Daugherty, A and LA Cassis (2002). Mechanisms of abdominal aortic aneurysm formation. *Current Atherosclerosis Reports*, 4, 222–227.
41. Davis, FM, DL Rateri and A Daugherty (2014). Mechanisms of aortic aneurysm formation: translating preclinical studies into clinical therapies. *Heart*, 100, 1498–1505.
42. Davis, FM, DL Rateri and A Daugherty (2015). Abdominal aortic aneurysm: novel mechanisms and therapies. *Current Opinion in Cardiology*, 30, 566–573.
43. Golledge, J, J Muller, A Daugherty and P Norman (2006). Abdominal aortic aneurysm: pathogenesis and implications for management. *Arteriosclerosis and Thrombosis: A Journal of Vascular Biology*, 26, 2605–2613.
44. Shimizu, K, RN Mitchell and P Libby (2006). Inflammation and cellular immune responses in abdominal aortic aneurysms. *Arteriosclerosis and Thrombosis: A Journal of Vascular Biology*, 26, 987–994.
45. Shimizu, K, M Shichiri, P Libby, RT Lee and RN Mitchell (2004). Th2-predominant inflammation and blockade of IFN-gamma signaling induce aneurysms in allografted aortas. *Journal of Clinical Investigation*, 114, 300–308.
46. Schonbeck, U, GK Sukhova, N Gerdes and P Libby (2002). T(H)2 predominant immune responses prevail in human abdominal aortic aneurysm. *American Journal of Pathology*, 161, 499–506.
47. Shimizu, K, P Libby and RN Mitchell (2005). Local cytokine environments drive aneurysm formation in allografted aortas. *Trends in Cardiovascular Medicine*, 15, 142–148.
48. Sun, J, GK Sukhova, J Zhang, H Chen, S Sjöberg, P Libby, M Xia, N Xiong, BD Gelb and G-P Shi (2012). Cathepsin K deficiency

reduces elastase perfusion-induced abdominal aortic aneurysms in mice. *Arteriosclerosis and Thrombosis: A Journal of Vascular Biology*, 32, 15–23.
49. Sun, J, GK Sukhova, J Zhang, H Chen, S Sjöberg, P Libby, M Xiang, J Wang, C Peters, T Reinheckel and G-P Shi (2011). Cathepsin L activity is essential to elastase perfusion-induced abdominal aortic aneurysms in mice. *Arteriosclerosis and Thrombosis: A Journal of Vascular Biology*, 31, 2500–2508.
50. Zhang, J, J Sun, JS Lindholt, GK Sukhova, M Sinnamon, RL Stevens, R Adachi, P Libby, RW Thompson and G-P Shi (2011). Mast cell tryptase deficiency attenuates mouse abdominal aortic aneurysm formation. *Circulation Research*, 108, 1316–1327.
51. Opie, LH, PJ Commerford, BJ Gersh and MA Pfeffer (2006). Controversies in ventricular remodelling. *Lancet*, 367, 356–367.
52. Libby, P, M Nahrendorf and FK Swirski (2016). Leukocytes link local and systemic inflammation in ischemic cardiovascular disease. *Journal of the American College of Cardiology*, 67, 1091–1103.
53. Lindsey, ML and R Zamilpa (2012). Temporal and spatial expression of matrix metalloproteinases and tissue inhibitors of metalloproteinases following myocardial infarction. *Cardiovascular Therapy*, 30, 31–41.
54. De Jesus, NM, L Wang, AW Herren, J Wang, F Shenasa, DM Bers, ML Lindsey and CM Ripplinger (2015). Atherosclerosis exacerbates arrhythmia following myocardial infarction: role of myocardial inflammation. *Heart Rhythm: The Official Journal of the Heart Rhythm Society*, 12, 169–178.
55. Chiao, YA, R Zamilpa, EF Lopez, Q Dai, GP Escobar, KW Hakala, ST Weintraub and ML Lindsey (2010). In vivo matrix metalloproteinase-7 substrates identified in the left ventricle post-Myocardial infarction using proteomics. *Journal of Proteome Research*, 9, 2649–2657.
56. Lindsey, ML, FA Zouein, Y Tian, R Padmanabhan Iyer and LE de Castro Bras (2015). Osteopontin is proteolytically processed by matrix metalloproteinase 9. *Canadian Journal of Physiology and Pharmacology*, 93, 879–886.
57. Gu, Z, J Cui, S Brown, R Fridman, S Mobashery, AY Strongin and SA Lipton (2005). A highly specific inhibitor of matrix metalloproteinase-9 rescues laminin from proteolysis and neurons from apoptosis

in transient focal cerebral ischemia. *Journal of Neuroscience*, 25, 6401–6408.
58. Zamilpa, R, EF Lopez, YA Chiao, Q Dai, GP Escobar, K Hakala, ST Weintraub and ML Lindsey (2010). Proteomic analysis identifies in vivo candidate matrix metalloproteinase-9 substrates in the left ventricle post-myocardial infarction. *Proteomics*, 10, 2214–2223.
59. Ma, Y, GV Halade, J Zhang, TA Ramirez, D Levin, A Voorhees, YF Jin, HC Han, AM Manicone and ML Lindsey (2013). Matrix metalloproteinase-28 deletion exacerbates cardiac dysfunction and rupture after myocardial infarction in mice by inhibiting m2 macrophage activation. *Circulation Research*, 112, 675–688.
60. Iyer, RP, NL Patterson, FA Zouein, Y Ma, V Dive, LE de Castro Bras and ML Lindsey (2015). Early matrix metalloproteinase-12 inhibition worsens post-myocardial infarction cardiac dysfunction by delaying inflammation resolution. *International Journal of Cardiology*, 185, 198–208.
61. Ma, Y, A Yabluchanskiy and ML Lindsey (2013). Neutrophil roles in left ventricular remodeling following myocardial infarction. *Fibrogenesis & Tissue Repair*, 6, 11.
62. Lambert, JM, EF Lopez and ML Lindsey (2008). Macrophage roles following myocardial infarction. *International Journal of Cardiology*, 130, 147–158.
63. Ma, Y, YA Chiao, J Zhang, AM Manicone, YF Jin and ML Lindsey (2012). Matrix metalloproteinase-28 deletion amplifies inflammatory and extracellular matrix responses to cardiac aging. *Microscopy and Microanalysis: The Official Journal of Microscopy Society of America, Microbeam Analysis Society, Microscopical Society of Canada*, 18, 81–90.
64. Porter, KE and NA Turner (2009). Cardiac fibroblasts: at the heart of myocardial remodeling. *Pharmacology & Therapeutics*, 123, 255–278.
65. Hughes, BG, X Fan, WJ Cho and R Schulz (2014). MMP-2 is localized to the mitochondria-associated membrane of the heart. *American Journal of Physiology Heart Circulatory Physiology*, 306, H764–H770.
66. Ali, MA and R Schulz (2009). Activation of MMP-2 as a key event in oxidative stress injury to the heart. *Frontiers in Bioscience*, 14, 699–716.

67. Matsumura, S-i, S Iwanaga, S Mochizuki, H Okamoto, S Ogawa and Y Okada (2005). Targeted deletion or pharmacological inhibition of MMP-2 prevents cardiac rupture after myocardial infarction in mice. *Journal of Clinical Investigation*, 115, 599–609.
68. Lovett, DH, R Mahimkar, RL Raffai, L Cape, BQ Zhu, ZQ Jin, AJ Baker and JS Karliner (2013). N-terminal truncated intracellular matrix metalloproteinase-2 induces cardiomyocyte hypertrophy, inflammation and systolic heart failure. *PLoS One*, 8, e68154.
69. Turner, NA, A Das, P Warburton, DJ O'Regan, SG Ball and KE Porter (2009). Interleukin-1alpha stimulates proinflammatory cytokine expression in human cardiac myofibroblasts. *American Journal of Physiology, Heart and Circulatory Physiology*, 297, H1117–H1127.
70. van Nieuwenhoven, FA, KE Hemmings, KE Porter and NA Turner (2013). Combined effects of interleukin-1alpha and transforming growth factor-beta1 on modulation of human cardiac fibroblast function. *Matrix Biology: Journal of the International Society for Matrix Biology*, 32(7–8), 399–406.
71. Lovelock, JD, AH Baker, F Gao, JF Dong, AL Bergeron, W McPheat, N Sivasubramanian and DL Mann (2005). Heterogeneous effects of tissue inhibitors of matrix metalloproteinases on cardiac fibroblasts. *American Journal of Physiology, Heart and Circulatory Physiology*, 288, H461–H468.
72. Sun, M, MA Opavsky, DJ Stewart, M Rabinovitch, F Dawood, W-H Wen and PP Liu (2003). Temporal response and localization of integrins {beta}1 and {beta}3 in the heart after myocardial infarction: regulation by cytokines. *Circulation*, 107, 1046–1052.
73. Starr, AE, A Dufour, J Maier and CM Overall (2012). Biochemical analysis of matrix metalloproteinase activation of chemokines CCL15 and CCL23 and increased glycosaminoglycan binding of CCL16. *Journal of Biological Chemistry*, 287, 5848–5860.
74. Van Den Steen, PE, A Wuyts, SJ Husson, P Proost, J Van Damme and G Opdenakker (2003). Gelatinase B/MMP-9 and neutrophil collagenase/MMP-8 process the chemokines human GCP-2/CXCL6, ENA-78/CXCL5 and mouse GCP-2/LIX and modulate their physiological activities. *European Journal of Biochemistry*, 270, 3739–3749.
75. Ramirez, TA, RP Iyer, O Ghasemi, EF Lopez, DB Levin, J Zhang, R Zamilpa, YM Chou, YF Jin and ML Lindsey (2014). Aliskiren and

valsartan mediate left ventricular remodeling post-myocardial infarction in mice through MMP-9 effects. *Journal of Molecular and Cellular Cardiology*, 72, 326–335.

76. Yabluchanskiy, A, Y Ma, RP Iyer, ME Hall and ML Lindsey (2013). Matrix metalloproteinase-9: many shades of function in cardiovascular disease. *Physiology (Bethesda)*, 28, 391–403.
77. Lindsey, M, K Wedin, MD Brown, C Keller, AJ Evans, J Smolen, AR Burns, RD Rossen, L Michael and M Entman (2001). Matrix-dependent mechanism of neutrophil-mediated release and activation of matrix metalloproteinase 9 in myocardial ischemia/reperfusion. *Circulation*, 103, 2181–2187.
78. Lindsey, ML, GP Escobar, LW Dobrucki, DK Goshorn, S Bouges, JT Mingoia, DM McClister Jr, H Su, J Gannon, C MacGillivray, RT Lee, AJ Sinusas and FG Spinale (2006). Matrix metalloproteinase-9 gene deletion facilitates angiogenesis after myocardial infarction. *American Journal of Physiology, Heart and Circulatory Physiology*, 290, H232–H239.
79. Zamilpa, R, JN Ibarra, LE de Castro Bras, TA Ramirez, N Nguyen, GV Halade, J Zhang, Q Dai, T Dayah, YA Chiao, W Lowell, SS Ahuja, J D'Armiento, YF Jin and ML Lindsey (2012). Transgenic overexpression of matrix metalloproteinase-9 in macrophages attenuates the inflammatory response and improves left ventricular function post-myocardial infarction. *Journal of Molecular Cell Cardiology*, 53, 599–608.
80. Tian, H, M Cimini, PW Fedak, S Altamentova, S Fazel, ML Huang, RD Weisel and RK Li (2007). TIMP-3 deficiency accelerates cardiac remodeling after myocardial infarction. *Journal of Molecular Cell Cardiology*, 43, 733–743.
81. Tian, H, ML Huang, KY Liu, ZB Jia, L Sun, SL Jiang, W Liu, HY Kinkaid, J Wu and RK Li (2011). Inhibiting matrix metalloproteinase by cell-based TIMP-3 gene transfer effectively treats acute and chronic ischemic cardiomyopathy. *Cell Transplantation*, 21, 1039–1053.
82. Polyakova, V, I Loeffler, S Hein, S Miyagawa, I Piotrowska, S Dammer, J Risteli, J Schaper and S Kostin (2011). Fibrosis in end-stage human heart failure: severe changes in collagen metabolism and MMP/TIMP profiles. *International Journal of Cardiology*, 151, 18–33.
83. Yang, DC, ST Ma, Y Tan, YH Chen, D Li, B Tang, JS Chen, XH Su, G Li, X Zhang and YJ Yang (2010). Imbalance of matrix

metalloproteinases/tissue inhibitor of metalloproteinase-1 and loss of fibronectin expression in patients with congestive heart failure. *Cardiology*, 116, 133–141.

84. Roger, VL, AS Go, DM Lloyd-Jones, RJ Adams, JD Berry, TM Brown, MR Carnethon, S Dai, G de Simone, ES Ford, CS Fox, HJ Fullerton, C Gillespie, KJ Greenlund, SM Hailpern, JA Heit, PM Ho, VJ Howard, BM Kissela, SJ Kittner, DT Lackland, JH Lichtman, LD Lisabeth, DM Makuc, GM Marcus, A Marelli, DB Matchar, MM McDermott, JB Meigs, CS Moy, D Mozaffarian, ME Mussolino, G Nichol, NP Paynter, WD Rosamond, PD Sorlie, RS Stafford, TN Turan, MB Turner, ND Wong and J Wylie-Rosett (2011). Heart disease and stroke statistics—2011 update: a report from the American Heart Association. *Circulation*, 123, e18–e209.
85. Solomon, SD, MSJ Sutton, GA Lamas, T Plappert, JL Rouleau, H Skali, L Moye, E Braunwald, MA Pfeffer and for the Survival and Ventricular Enlargement (SAVE) Investigators (2002). Ventricular remodeling does not accompany the development of heart failure in diabetic patients after myocardial infarction. *Circulation*, 106, 1251–1255.
86. Lavine, SJ (2003). Prediction of heart failure post myocardial infarction: comparison of ejection fraction, transmitral filling parameters, and the index of myocardial performance. *Echocardiography*, 20, 691–701.
87. Hellermann, JP, SJ Jacobsen, MM Redfield, GS Reeder, SA Weston and VL Roger (2005). Heart failure after myocardial infarction: clinical presentation and survival. *European Journal of Heart Failure*, 7, 119–125.
88. Hellermann, JP, TY Goraya, SJ Jacobsen, SA Weston, GS Reeder, BJ Gersh, MM Redfield, RJ Rodeheffer, BP Yawn and VL Roger (2003). Incidence of heart failure after myocardial infarction: is it changing over time? *American Journal of Epidemiology*, 157, 1101–1107.
89. Anavekar, NS, JJV McMurray, EJ Velazquez, SD Solomon, L Kober, J-L Rouleau, HD White, R Nordlander, A Maggioni, K Dickstein, S Zelenkofske, JD Leimberger, RM Califf and MA Pfeffer (2004). Relation between renal dysfunction and cardiovascular outcomes after myocardial infarction. *New England Journal of Medicine*, 351, 1285–1295.
90. Sutton, MSJ, MA Pfeffer, L Moye, T Plappert, JL Rouleau, G Lamas, J Rouleau, JO Parker, MO Arnold, B Sussex and E Braunwald (1997).

Cardiovascular death and left ventricular remodeling two years after myocardial infarction: baseline predictors and impact of long-term use of captopril: information from the survival and ventricular enlargement (SAVE) trial. *Circulation*, 96, 3294–3299.

91. Ezekowitz ,JA, P Kaul, JA Bakal, PW Armstrong, RC Welsh and FA McAlister (2009). Declining in-hospital mortality and increasing heart failure incidence in elderly patients with first myocardial infarction. *Journal of the American College of Cardiology*, 53, 13–20.
92. de Denus, S, J Lavoie, A Ducharme, E O'Meara, N Racine, MG Sirois, PE Neagoe, L Zhu, JL Rouleau and M White (2012). Differences in biomarkers in patients with heart failure with a reduced vs a preserved left ventricular ejection fraction. *Canadian Journal of Cardiology*, 28, 62–68.
93. Spinale, FG and NM Wilbur (2009). Matrix metalloproteinase therapy in heart failure. *Current Treatment Options in Cardiovascular Medicine*, 11, 339–346.
94. Messerli, FH (2004). TIMPs, MMPs and cardiovascular disease. *European Heart Journal,* 25, 1475–1476.
95. Sundstrom, J, JC Evans, EJ Benjamin, D Levy, MG Larson, DB Sawyer, DA Siwik, WS Colucci, P Sutherland, PW Wilson and RS Vasan (2004a). Relations of plasma matrix metalloproteinase-9 to clinical cardiovascular risk factors and echocardiographic left ventricular measures: the Framingham Heart Study. *Circulation*, 109, 2850–2856.
96. Sundstrom, J, JC Evans, EJ Benjamin, D Levy, MG Larson, DB Sawyer, DA Siwik, WS Colucci, PW Wilson and RS Vasan (2004b). Relations of plasma total TIMP-1 levels to cardiovascular risk factors and echocardiographic measures: the Framingham heart study. *European Heart Journal*, 25, 1509–1516.
97. Sundstrom J and RS Vasan (2006). Circulating biomarkers of extracellular matrix remodeling and risk of atherosclerotic events. *Current Opinion in Lipidology,* 17, 45–53.
98. Velagaleti RS, P Gona, J Sundstrom, MG Larson, D Siwik, WS Colucci, EJ Benjamin and RS Vasan (2010). Relations of biomarkers of extracellular matrix remodeling to incident cardiovascular events and mortality. *Arteriosclerosis and Thrombosis: A Journal of Vascular Biology,* 30, 2283–2288.
99. Hansson, J, L Lind, J Hulthe and J Sundstrom (2009). Relations of serum MMP-9 and TIMP-1 levels to left ventricular measures and

cardiovascular risk factors: a population-based study. *European Journal of Cardiovascular Prevention and Rehabilitation: Official Journal of the European Society of Cardiology, Working Groups on Epidemiology & Prevention and Cardiac Rehabilitation and Exercise Physiology*, 16, 297–303.

100. Hansson, J, RS Vasan, J Arnlov, E Ingelsson, L Lind, A Larsson, K Michaelsson and J Sundstrom (2011). Biomarkers of extracellular matrix metabolism (MMP-9 and TIMP-1) and risk of stroke, myocardial infarction, and cause-specific mortality: cohort study. *PLoS One*, 6, e16185.
101. Blankenberg, S, HJ Rupprecht, O Poirier, C Bickel, M Smieja, G Hafner, J Meyer, F Cambien and L Tiret (2003). Plasma concentrations and genetic variation of matrix metalloproteinase 9 and prognosis of patients with cardiovascular disease. *Circulation*, 107, 1579–1585.
102. Lubos, E, R Schnabel, HJ Rupprecht, C Bickel, CM Messow, S Prigge, F Cambien, L Tiret, T Munzel and S Blankenberg (2006). Prognostic value of tissue inhibitor of metalloproteinase-1 for cardiovascular death among patients with cardiovascular disease: results from the AtheroGene study. *European Heart Journal*, 27, 150–156.
103. Deleon-Pennell, KY, R Altara, A Yabluchanskiy, A Modesti and ML Lindsey (2015). The circular relationship between matrix metalloproteinase-9 and inflammation following myocardial infarction. *International Union of Biochemistry and Molecular Biology Life*, 67, 611–618.
104. Iyer, RP, LE de Castro Bras, YF Jin and ML Lindsey (2014). Translating Koch's postulates to identify matrix metalloproteinase roles in postmyocardial infarction remodeling: cardiac metalloproteinase actions (CarMA) postulates. *Circulation Research*, 114, 860–871.
105. Yabluchanskiy, A, Y Li, RJ Chilton and ML Lindsey (2013). Matrix metalloproteinases: drug targets for myocardial infarction. *Current Drug Targets*, 14, 276–286.
106. Halade, GV, YF Jin and ML Lindsey (2013). Matrix metalloproteinase (MMP)-9: a proximal biomarker for cardiac remodeling and a distal biomarker for inflammation. *Pharmacology & Therapeutics*, 139, 32–40.
107. Iyer, RP, NL Patterson, GB Fields and ML Lindsey (2012). The history of matrix metalloproteinases: milestones, myths, and

misperceptions. *American Journal of Physiology Heart and Circulatory Physiology*, 303, H919–H930.

108. Fisher, J and S Mobashery (2006). Recent advances in MMP inhibitor design. *Cancer Metastasis Reviews*, 25, 115–136.
109. Marchant, DJ, CL Bellac, TJ Moraes, SJ Wadsworth, A Dufour, GS Butler, LM Bilawchuk, RG Hendry, AG Robertson, CT Cheung, J Ng, L Ang, Z Luo, K Heilbron, MJ Norris, W Duan, T Bucyk, A Karpov, L Devel, D Georgiadis, RG Hegele, H Luo, DJ Granville, V Dive, BM McManus and CM Overall (2014). A new transcriptional role for matrix metalloproteinase-12 in antiviral immunity. *Nature Medicine*, 20, 493–502.
110. Johnson, JL, L Devel, B Czarny, SJ George, CL Jackson, V Rogakos, F Beau, A Yiotakis, AC Newby and V Dive (2011). A selective matrix metalloproteinase-12 inhibitor retards atherosclerotic plaque development in apolipoprotein E-knockout mice. *Arteriosclerosis, Thrombosis, and Vascular Biology*, 31, 528–535.
111. Devel, L, B Czarny, F Beau, D Georgiadis, E Stura and V Dive (2010). Third generation of matrix metalloprotease inhibitors: gain in selectivity by targeting the depth of the S1' cavity. *Biochimie*, 92, 1501–1508.
112. Lauer-Fields, JL and GB Fields (2009). Development of selective, exosite binding matrix metalloproteinase (MMP) inhibitors. In *Peptides: Breaking Away (Proceedings of the Twenty-First American Peptide Symposium)*, M Lebl (ed.), pp. 239–241. Albuquerque, NM: American Peptide Society.
113. Lauer-Fields, JL, JK Whitehead, S Li, RP Hammer, K Brew and GB Fields (2008). Selective modulation of matrix metalloproteinase 9 (MMP-9) functions via exosite inhibition. *Journal of Biological Chemistry*, 283, 20087–20095.
114. Lamblin, N, C Bauters, X Hermant, JM Lablanche, N Helbecque and P Amouyel (2002). Polymorphisms in the promoter regions of MMP-2, MMP-3, MMP-9 and MMP-12 genes as determinants of aneurysmal coronary artery disease. *Journal of the American College of Cardiology*, 40, 43–48.
115. Weber, C, A Zernecke and P Libby (2008). The multifaceted contributions of leukocyte subsets to atherosclerosis: lessons from mouse models. *Nature Reviews Immunology*, 8, 802–815.
116. Hansson, GK, P Libby and I Tabas (2015). Inflammation and plaque vulnerability. *Journal of Internal Medicine*, 278, 483–493.

117. Libby, P, AH Lichtman and GK Hansson (2013). Immune effector mechanisms implicated in atherosclerosis: from mice to humans. *Immunity*, 38, 1092–1104.
118. Libby, P (2012). Inflammation in atherosclerosis. *Arteriosclerosis, Thrombosis, and Vascular Biology*, 32, 2045–2051.
119. Koullias, GJ, P Ravichandran, DP Korkolis, DL Rimm and JA Elefteriades (2004). Increased tissue microarray matrix metalloproteinase expression favors proteolysis in thoracic aortic aneurysms and dissections. *Annals of Thoracic Surgery*, 78, 2106–2110.
120. Wilson, WRW, M Anderton, EC Schwalbe, JL Jones, PN Furness, PRF Bell and MM Thompson (2006). Matrix metalloproteinase-8 and-9 are increased at the site of abdominal aortic aneurysm rupture. *Circulation*, 113, 438–445.
121. Spinale, FG and F Villarreal (2014). Targeting matrix metalloproteinases in heart disease: lessons from endogenous inhibitors. *Biochemical Pharmacology*, 90, 7–15.
122. Peterson, JT (2006). The importance of estimating the therapeutic index in the development of matrix metalloproteinase inhibitors. *Cardiovascular Research*, 69, 677–687.
123. Murphy, G (2008). The ADAMs: signalling scissors in the tumour microenvironment. *Nature Reviews Cancer*, 8, 932–941.
124. Hudson, MP, PW Armstrong, W Ruzyllo, J Brum, L Cusmano, P Krzeski, R Lyon, M Quinones, P Theroux, D Sydlowski, HE Kim, MJ Garcia, WA Jaber and WD Weaver (2006). Effects of selective matrix metalloproteinase inhibitor (PG-116800) to prevent ventricular remodeling after myocardial infarction: results of the PREMIER (Prevention of Myocardial Infarction Early Remodeling) trial. *Journal of the American College of Cardiology*, 48, 15–20.
125. Dorman, G, S Cseh, I Hajdu, L Barna, D Konya, K Kupai, L Kovacs and P Ferdinandy (2010). Matrix metalloproteinase inhibitors: a critical appraisal of design principles and proposed therapeutic utility. *Drugs*, 70, 949–964.
126. Bench, TJ, A Jeremias and DL Brown (2011). Matrix metalloproteinase inhibition with tetracyclines for the treatment of coronary artery disease. *Pharmacological Research: The Official Journal of the Italian Pharmacological Society*, 64, 561–566.
127. Benjamin, MM and RA Khalil (2012). Matrix metalloproteinase inhibitors as investigative tools in the pathogenesis and management of vascular disease. *EXS*, 103, 209–279.

128. Golub, LM, S Ciancio, NS Tamamurthy, M Leung and TF McNamara (1990). Low-dose doxycycline therapy: effect of gingival and crevicular fluid collagenase activity in humans. *Journal of Periodontal Research*, 25, 321–330.
129. Karimbux, NY, NS Ramamurthy, LM Golub and I Nishimura (1994). Tissue healing with doxycycline and chemically modified tetracycline treatments in rats with porphyromonas gingivalis-induced periodontitis. *Annals of the New York Academy of Sciences*, 732, 433–435.
130. Villarreal, FJ, M Griffin, J Omens, W Dillmann, J Nguyen and J Covell (2003). Early short-term treatment with doxycycline modulates postinfarction left ventricular remodeling. *Circulation*, 108, 1487–1492.
131. Courtman, DW, CD Franco, Q Meng and MP Bendeck (2004). Inward remodeling of the rabbit aorta is blocked by the matrix metalloproteinase inhibitor doxycycline. *Journal of Vascular Research*, 41, 157–165.
132. Garcia, RA, KV Go and FJ Villarreal (2007). Effects of timed administration of doxycycline or methylprednisolone on post-myocardial infarction inflammation and left ventricular remodeling in the rat heart. *Molecular and Cell Biochemistry*, 300, 159–169.
133. Lindeman, JHN, H Abdul-Hussien, JH van Bockel, R Wolterbeek and R Kleemann (2009). Clinical trial of doxycycline for matrix metalloproteinase-9 inhibition in patients with an abdominal aneurysm: doxycycline selectively depletes aortic wall neutrophils and cytotoxic T cells. *Circulation*, 119, 2209–2216.
134. Cerisano, G, P Buonamici, R Valenti, G Moschi, E Taddeucci, L Giurlani, A Migliorini, R Vergara, G Parodi, R Sciagra, R Romito, P Colonna and D Antoniucci (2014). Effects of a timely therapy with doxycycline on the left ventricular remodeling according to the pre-procedural TIMI flow grade in patients with ST-elevation acute myocardial infarction. *Basic Research in Cardiology*, 109, 412.
135. Cerisano, G, P Buonamici, R Valenti, R Sciagra, S Raspanti, A Santini, N Carrabba, EV Dovellini, R Romito, A Pupi, P Colonna and D Antoniucci (2014). Early short-term doxycycline therapy in patients with acute myocardial infarction and left ventricular dysfunction to prevent the ominous progression to adverse remodelling: the TIPTOP trial. *European Heart Journal*, 35, 184–191.
136. Death, AK, S Nakhla, KC McGrath, S Martell, DK Yue, W Jessup and DS Celermajer (2002). Nitroglycerin upregulates matrix

metalloproteinase expression by human macrophages. *Journal of the American College of Cardiology*, 39, 1943–1950.
137. Cuadrado, E, L Ortega, M Hernandez-Guillamon, A Penalba, I Fernandez-Cadenas, A Rosell and J Montaner (2008). Tissue plasminogen activator (t-PA) promotes neutrophil degranulation and MMP-9 release. *Journal of Leukocyte Biology*, 84, 207–214.
138. Tziakas, DN, GK Chalikias, EI Hatzinikolaou, DA Stakos, IK Tentes, A Kortsaris, DI Hatseras and JC Kaski (2006). Alteplase treatment affects circulating matrix metalloproteinase concentrations in patients with ST segment elevation acute myocardial infarction. *Thrombosis Research*, 118, 221–227.
139. Rababah, M, H Worthmann, M Deb, AB Tryc, YT Ma, OM El Bendary, H Hecker, A Goldbecker, M Heeren, K Brand, K Weissenborn and R Lichtinghagen (2012). Anticoagulants affect matrix metalloproteinase 9 levels in blood samples of stroke patients and healthy controls. *Clinical Biochemistry*, 45, 483–489.
140. Castellazzi, M, C Tamborino, E Fainardi, MC Manfrinato, E Granieri, F Dallocchio and T Bellini (2007). Effects of anticoagulants on the activity of gelatinases. *Clinical Biochemistry*, 40, 1272–1276.
141. Reinhardt, D, HH Sigusch, J Hensse, SC Tyagi, R Korfer and HR Figulla (2002). Cardiac remodelling in end stage heart failure: upregulation of matrix metalloproteinase (MMP) irrespective of the underlying disease, and evidence for a direct inhibitory effect of ACE inhibitors on MMP. *Heart*, 88, 525–530.
142. McElmurray, JH, 3rd, R Mukherjee RB, New, AC Sampson, MK, King JW Hendrick, A Goldberg, TJ Peterson, H Hallak, MR Zile and FG Spinale (1999). Angiotensin-converting enzyme and matrix metalloproteinase inhibition with developing heart failure: comparative effects on left ventricular function and geometry. *Journal of Pharmacology and Experimental Therapeutics*, 291, 799–811.
143. Yamamoto, D, S Takai, D Jin, S Inagaki, K Tanaka and M Miyazaki (2007). Molecular mechanism of imidapril for cardiovascular protection via inhibition of MMP-9. *Journal of Molecular and Cellular Cardiology*, 43, 670–676.
144. Yamamoto, D, S Takai and M Miyazaki (2007). Prediction of interaction mode between a typical ACE inhibitor and MMP-9 active site. *Biochemical and Biophysical Research Communications*, 354, 981–984.

145. Yamamoto, D, S Takai and M Miyazaki (2008). Inhibitory profiles of captopril on matrix metalloproteinase-9 activity. *European Journal of Pharmacology*, 588, 277–279.
146. Yamamoto, D and S Takai (2009). Pharmacological implications of MMP-9 inhibition by ACE inhibitors. *Current Medical Chemistry*, 16, 1349–1354.
147. Jin, Y, HC Han and ML Lindsey (2007). ACE inhibitors to block MMP-9 activity: new functions for old inhibitors. *Journal of Molecular and Cellular Cardiology*, 43, 664–666.
148. Gajarsa, JJ and RA Kloner (2011). Left ventricular remodeling in the post-infarction heart: a review of cellular, molecular mechanisms, and therapeutic modalities. *Heart Failure Reviews*, 16, 13–21.
149. Tziakas, DN, GK Chalikias, JT Parissis, EI Hatzinikolaou, ED Papadopoulos, GA Tripsiannis, EG Papadopoulou, IK Tentes, SM Karas and DI Chatseras (2004). Serum profiles of matrix metalloproteinases and their tissue inhibitor in patients with acute coronary syndromes. The effects of short-term atorvastatin administration. *International Journal of Cardiology*, 94, 269–277.
150. Turner, NA, PK Aley, KT Hall, P Warburton, S Galloway, L Midgley, DJ O'Regan, IC Wood, SG Ball and KE Porter (2007). Simvastatin inhibits TNFalpha-induced invasion of human cardiac myofibroblasts via both MMP-9-dependent and -independent mechanisms. *Journal of Molecular and Cellular Cardiology*, 43, 168–176.
151. Luan, Z, AJ Chase and AC Newby (2003). Statins inhibit secretion of metalloproteinases-1,-2,-3, and-9 from vascular smooth muscle cells and macrophages. *Arteriosclerosis and Thrombosis: A Journal of Vascular Biology*, 23, 769–775.
152. Yasuda, S, S Miyazaki, H Kinoshita, N Nagaya, M Kanda, Y Goto and H Nonogi (2007). Enhanced cardiac production of matrix metalloproteinase-2 and-9 and its attenuation associated with pravastatin treatment in patients with acute myocardial infarction. *Clinical Science (Lond)*, 112, 43–49.
153. Koh, KK, JW Son, JY Ahn, DK Jin, HS Kim, YM Choi, DS Kim, EM Jeong, GS Park, IS Choi and EK Shin (2002). Comparative effects of diet and statin on NO bioactivity and matrix metalloproteinases in hypercholesterolemic patients with coronary artery disease. *Arteriosclerosis, Thrombosis, and Vascular Biology*, 22, e19–e23.

154. Adamek, A, K Hu, B Bayer, H Wagner, G Ertl, J Bauersachs and S Frantz (2007). High dose aspirin and left ventricular remodeling after myocardial infarction: aspirin and myocardial infarction. *Basic Research in Cardiology*, 102, 334–340.
155. Pan, M-R and W-C Hung (2002). Nonsteroidal anti-inflammatory drugs inhibit matrix metalloproteinase-2 via suppression of the ERK/Sp1-mediated transcription. *Journal of Biological Chemistry*, 277, 32775–32780.
156. Haffner, SM, AS Greenberg, WM Weston, H Chen, K Williams and MI Freed (2002). Effect of rosiglitazone treatment on nontraditional markers of cardiovascular disease in patients with type 2 diabetes mellitus. *Circulation*, 106, 679–684.
157. Eickelberg, O, M Roth, R Mussmann, JJ Rudiger, M Tamm, AP Perruchoud and LH Block (1999). Calcium channel blockers activate the interleukin-6 gene via the transcription factors NF-IL6 and NF-kappaB in primary human vascular smooth muscle cells. *Circulation*, 99, 2276–2282.

9. Matrix Metalloproteinases (MMPs) as Therapeutic Targets

Constance E. Brinckerhoff

9.1 Introduction: Understanding the Failure of Early Clinical Trials: "Obstacles to the Therapeutic use of MMP Inhibitors"

Blocking the heightened levels of matrix metalloproteinase (MMP) activity associated with disease has long been a therapeutic goal,[1–4] either by inhibiting enzymatic activity with small molecule chemical inhibitors or by blocking expression of the MMP genes. However, seemingly constant "obstacles to the therapeutic use of MMP inhibitors"[2] related to issues of specificity have plagued the field, and efficacy has been consistently disappointing (Table 9.1).[2] In addition, the disastrous and unexpected side effects emerging from the clinical trials at the end of the 20th century seemed to put the final nail in the "therapeutic coffin". Now, as we understand more and more about the diverse biology of these enzymes, interest in targeting MMPs therapeutically is reviving, but with a much more sophisticated approach.

Fifteen or 20 years ago, the concept that MMPs had functions in addition to degrading matrix proteins was still in its infancy. Furthermore, it was well accepted that rampant over-expression of MMPs was largely responsible for the disease pathology associated with arthritis, cancer and cardiovascular disease, and perhaps other

Table 9.1. Obstacles to the therapeutic use of MMP inhibitors.

Despite promising preclinical data supporting the blockade of matrix metalloproteinases (MMPs) as a treatment for cancer more than 10 years ago, all Phase III cancer trails failed owing to several reasons, as outlined below:

- The design of the clinical trails was suboptimal and inadequate clinical end points were used.
- The MMP inhibitors that were used were not ideal. Several were metabolically unstable, poorly orally bioavailable and/or associated with a dose-limiting toxicity. Moreover, early clinical trials with broad-spectrum MMP inhibitors revealed that prolonged treatment was associated with unwanted side effects and this resulted in the use of lower, often inadequate, MMP inhibitor doses in subsequent trials.
- At that time, there was poor knowledge on the complexity of MMP function and not every MMP had been identified. Consequently, inhibition of antitarget MMPs was not prevented and board-spectrum MMP inhibitors were used, not taking into account time- and space-specific, and sometimes opposing, MMP functions. Indeed, the undesired side effects were most probably due to the board-spectrum inhibition of MMPs, as well as the cross-inhibition of members of the adisintegrin and metalloproteinase (ADAM) family and aggrecanases (ADAMs with thrombospondin motifs (ADAMTS) family members).
- Finally, the discrepancies between the mouse models used in preclinical studies and the patients enrolled in the clinical trails likely contributed to the failure of the clinical trails. Indeed, patients were often in advanced stages of disease, whereas the animal studies had shown effectiveness of MMP inhibitors in early stages of disease.

Over the past two decades, the family of MMPs has been studied in more detail in several mammalian species, both at the gene and protein level, in health and disease. Many MMP-coding genes have been knocked out in mouse models, providing an *in vivo* system to investigate the consequences of the absence of these genes. Based on this research, MMPs are still regarded as important biological mediators that are detrimental in several diseases. It is intriguing why, despite their targetability, the development and marketing of specific MMP inhibitors has been delayed so much. Here, we summarize several possible reasons.

- The determinants of substrate specificity of MMPs are not well understood. Unlike other proteases, such as caspases, most MMPs have no obvious or strict consensus amino acid recognition sequence. This hampers the production of substrate-based inhibitors.

(*Continued*)

Table 9.1. (*Continued*)

• There is uncertainity in translating *in vitro* substrate information to *in vivo* relevance. Incubating a biologically active MMP with another protein *in vitro* will often result in a specific cleavage. However, this cleavage does not necessarily occur in an *in vivo* setting during health or disease.
• Many MMPs have evolved through gene duplication in the mammalian genome, leading to clusters of MMP genes on particular chromosomes (for example, the proximal mouse chromosome 9 harboursten MMP genes in less than 500 kilobases) and to extensive homology in amino acid sequence. This impedes the development of specific MMP inhibitors and of reliable tools to detect specific MMP activity *in vivo*.
• MMP members have overlapping biological substrates, and therefore substrate-based inhibitors can rarely be specific.
• As most MMP-deficient mice do not show an obvious phenotypic abnormality in unstimulated conditions, except for MMP14 and MMP20-deficient mice, not much has been invested in tissue-specific knockout mice. Consequently, there is insufficient knowledge on the spatiot emporal activities of MMPs *in vivo* during development, health and pathological conditions.
• The first clinical trial failures in the use of MMP inhibitors for cancer caused a shock wave in the MMP research community and beyond and made biotechnology and pharmaceutical companies reluctant to invest in newer generation MMP inhibitors.
• The high level of conservation of MMP genes in mammals indicates that they are essential for normal functioning of the organism. Hence, not all biological activities of MMPs are harmful, and although they could function as therapeutic targets in some conditions, they may be antitargets in other conditions.

Source: From Vandenbroucke and Libert (2014).

maladies as well. Thus, the idea of inhibiting MMPs to reduce disease pathology seemed both obvious and full of potential to create a therapeutic "bang". And so, the era of developing MMP inhibitors (MMPIs) was born, and the focus was simply (and in hindsight, naïve) on developing broad-spectrum inhibitors of MMPs.[2,3,5] The success of preclinical studies with batimastat (also known as BB-94) in the early 1990s demonstrated potential as powerful anticancer agents, fueling excitement and interest in developing these compounds. Soon, more than 50 MMPIs were in various stages of testing in clinical trials. Imagine, then, the overwhelming dismay when all of the anticancer trials failed, despite the positive findings with preclinical testing. Many reasons have been suggested to explain these failures (Table 9.1)[2]: poor trial design, inadequate clinical end points, metabolic instability and poor oral bioavailability of the MMPIs, and dose-limiting toxicities. Given the increasing information that MMPs could cleave many nonmatrix proteins, many of which have essential functions in normal physiology and homeostasis, it is not surprising that the clinical trials with broad-based MMPIs were toxic, with painful and unacceptable musculoskeletal side-effects. Although these effects were reversible, subsequent trials reduced the dosage to try to avoid them, only to reach the point of inefficacy.[2]

However, most troubling were results of two trials involving the MMPI tanomastat, a so-called "second generation" (and supposedly improved) hydroxamate inhibitor, in small-cell lung cancer and pancreatic cancer, both of which were stopped prematurely when patients receiving them had significantly shorter survival compared to patients receiving placebo.[1,2] We now surmise that these broad-based MMPIs were blocking the activities of some MMPs that actually had beneficial anticancer effects. This realization gave birth to the concept that some MMPs should be regarded as "antitargets". Further, the undesirable side effects of musculo-skeletal pain were likely the results of broadly inhibiting many (if not all) MMPs, along with some cross-inhibition of the ADAMs and of ADAMTS (aggrecanases) family members. These enzymes, like the MMPs have important roles in cellular adhesion, cell–matrix interactions, the release and activation of growth factors from matrix, along with mediating the shedding of membrane-bound proteins. Another explanation for the failure of the clinical trials is the fact that patients in the trials were often in the advanced stages of disease, while in the preclinical studies with animals, the drugs showed efficacy in these model systems.

Despite these early disasters, new and more sophisticated understanding of MMP structure and function, along with the development of novel technologies renewed interest in developing

MMPIs with greater selectivity for individual MMPs. Mostly, this renewed interest has focused on inhibiting enzyme activity, rather than blocking gene expression. The hope is that this approach will lead to more targeted efficacy with fewer side-effects. However, success will depend on the knowledge of the biochemistry of each MMP, and on the protocols developed for using newly derived compounds.

9.2 Basic Biochemistry of MMPs and the History of MMPIs

The first generation of these inhibitors blocked the enzymatic activity of MMPs by binding to the Zn++ ion in their active sites. Along with this Zn++ ion, MMPs have subsites (S), which are labeled as unprimed (left-hand side of the Zn++ ion; S1, S2, and S3) or primed (right-hand side of the Zn++ ion; S1′, S2′ and S3′) (Fig. 9.1).[2,5] The S1′ pocket is the principal subsite for substrate recognition and is the most variable subsite among all the MMPs with respect to the amino acid sequence and the depth of the pocket. MMPs can be grouped, on the basis of the depth of this pocket, into shallow, intermediate and deep-pocket MMPs. The P1′–S1′ interaction (P1 is the group in

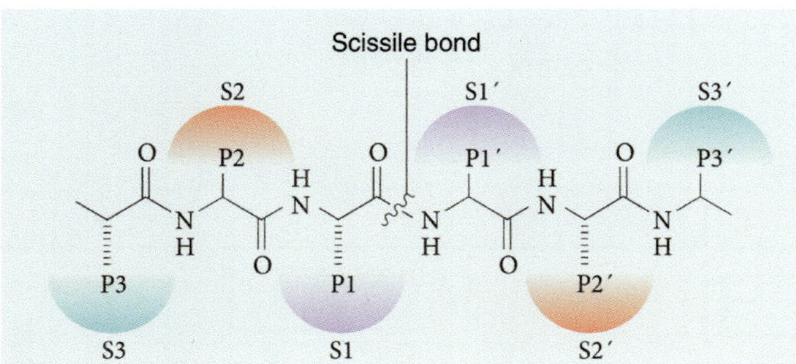

Figure 9.1. Nomenclature of the substrate–protease interaction. Multiple enzymatic binding sites (P) directly contact the substrate-binding sites (S) of the substrate. The nomenclature of the S sites (for example S1–3 and S1′–3′) is concordant with the P sites. Sn sites are amino-terminal to the scissile bond, and Sn′ sites are carboxy-terminal to the scissile bond. (From Vandenbroucke and Libert, 2014.)

the inhibitor or substrate that binds to the S1′ pocket of the enzyme) is the primary determinant of the affinity of inhibitors and the cleavage position of peptide substrates. For instance, extending the P1 substituent for MMP-13 resulted in increased selectivity over the very homologous MMP-2.[2,5,6]

The development of high throughput proteomic techniques and conditional genetic modifications of mouse models are facilitating our increased understanding of the complexities of MMP function and emphasizing the difficulties in developing specific inhibitors.[7] In addition, another hurdle for the development of specific inhibitors of catalysis has been the broad similarity of the MMP catalytic site (Figs. 9.2 and 9.3)[6,8] More detailed knowledge of the active site structure has helped in developing chemical inhibitors with greater specificity, and selected enzymes are now being targeted.[5] The role of exosites, that is, secondary binding sites, both in the catalytic and hemopexin domains, is being explored as determinants of substrate specificity of individual MMPs, and hence, as potential specific targeting sites for inhibitors.

Initial efforts to develop MMPIs were based on the structure of collagen.[2,6] The catalytic domains of MMPs have a shallow active

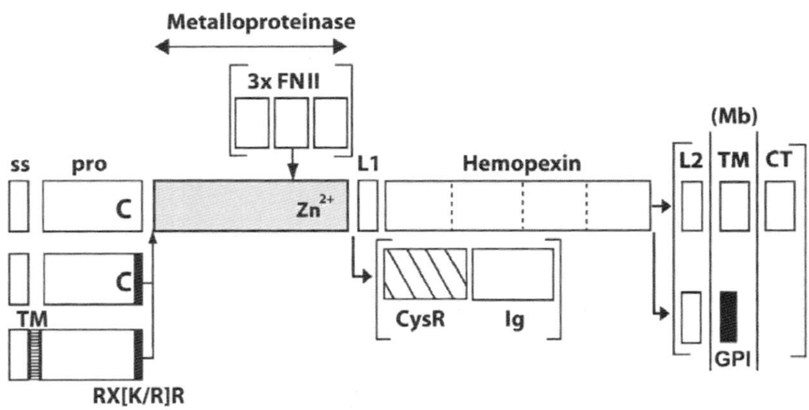

Figure 9.2. Domain structures of the MMP family. ss, signal sequence; pro, prodomain, FNII, fibronectin type II motif; L1, linker 1; L2, linker 2; Mb, plasma membrane; TM, transmembrane domain; CT, cytoplamic tail; CysR, cysteine rich; Ig, immunoglobulin domain; GPI, glycosylphosphatidylinositol anchor; C, cysteine. (From Murphy and Nagase, 2008.)

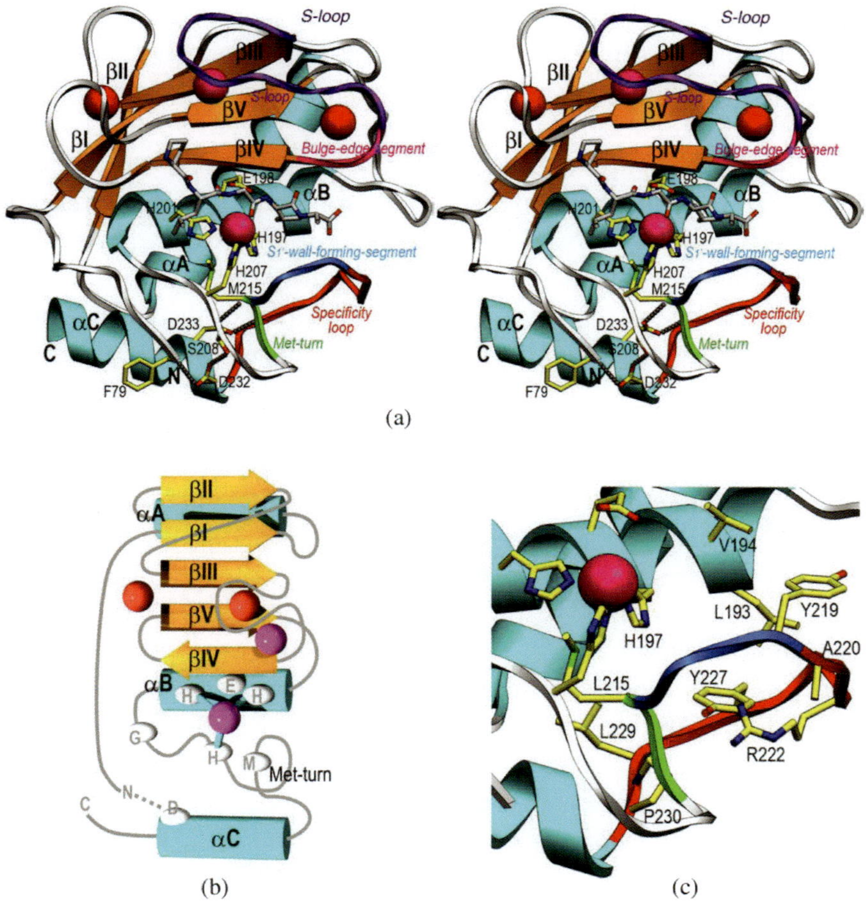

Figure 9.3. MMP catalytic domain structure. (a) Stereographic Richardson-plot of the catalytic domain of human MMP-8. The repetitive secondary structure elements (orange arrows for β-strands, βI–βV; cyan ribbons for α-helices, αA-αC) and the four cations (two zinc ions in magenta and two calcium ions in red) are depicted. The side chains of the zinc-binding histidines, the general base/acid glutamate, the Met-turn methionine, and residues engaged in key electrostatic interactions (grey dots) within the CTS are shown as stick models with yellow carbons and labeled. A substrate of sequence Pro-Leu-Gly-Leu-Ala, modeled based on published inhibitor structures is further shown as a stick model with grey carbons. Additional relevant chain segments are shown in distinct colors and labeled (Met-turn in green; specificity loop in red; S1′-wall-forming segment in blue; S-loop in purple; and bulge-edge segment in magenta). (b) Topology scheme of MMP-8 in the same orientation as in (a). (c) Close-up view of (a) depicting the side chains engaged in zinc binding and those shaping the specificity pocket, which are labeled. (From Tallant *et al.*, 2010.)

site cleft with a flat unprimed side and a narrow primed side centered around the S01 subsite (Fig. 9.1).[2,5] MMPs share catalytic domains of 160–170 residues with a marked sequence similarity, where the percentage of identical residues ranges from 33% to 90%. The first inhibitors were hydroxamate-based, and were made up of the basic backbone of collagen and the zinc-binding group hydroxamate (–CONHOH), and thus, they were named hydroxamate-based MMPIs (Fig. 9.4). These compounds contain a group that chelates the catalytic Zn++ ion and a backbone that mimics the natural peptide substrate of a particular MMP or group of MMPs.[2,5] Hydroxamate is a powerful ligand that binds to the catalytic Zn++ ion, resulting in the formation of a distorted trigonal–bipyramidal structure around the Zn++ ion. The –NH group of the hydroxamate anion hydrogen bonds with the neighboring carbonyl oxygen, and hydrophobic interactions then stabilize the inhibitor–enzyme complex. However, the hydoxymate Zn-binding group is metabolically labile, giving the drugs a relatively short half-life.

Along with batimastat, hydroxamate collagen-based peptidomimetic MMPIs include marimastat and ilomastat (also known as GM6001). Despite biochemical efficacy, the low water solubility of batismastat made drug delivery difficult for clinical trials.[1,2] Marimastat, on the other hand, was readily water soluble but not therapeutically effective at doses that caused musculoskeletal toxicity when administered in randomized Phase III trial for metastatic breast cancer.[2] Similarly, a Phase I trial of marimastat along with carboplatin and paclitaxel as a therapy against advanced non-small-cell lung cancer was also associated musculoskeletal toxicity, even though in this case, the patients did show a positive therapeutic response. Moreover, it became clear that clinical efficacy required that the MMPIs be administered simultaneously with the highest levels of MMP enzymatic activity, which in a particular cancer, was often early in the course of disease development, perhaps even before clinical detection/diagnosis. Thus, MMPIs were often given too late to be helpful.

Given the enthusiasm over MMPIs and their therapeutic potential, a second generation of hydroxamate-based MMPIs was engineered. They were composed of a substituted aryl, a sulphonamide and a

Inhibitor (alternative names)	Structure	MMP inhibition
Batimastat (BB-94)	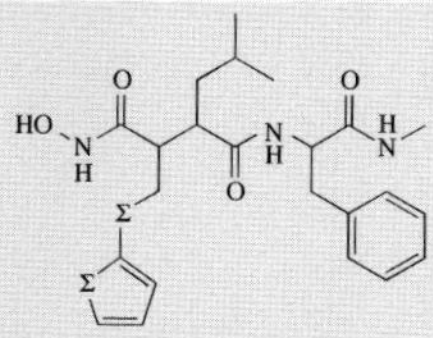	IC_{50} MMP1 = 3 nM IC_{50} MMP2 = 4 nM IC_{50} MMP3 = 20 nM IC_{50} MMP7 = 6 nM IC_{50} MMP8 = 10 nM IC_{50} MMP9 = 1 nM
BB-1101	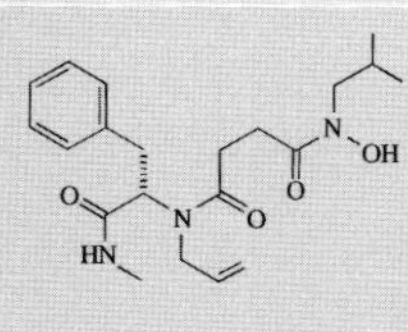	K_i MMP1 = 8 nM K_i MMP2 = 4 nM K_i MMP3 = 30 nM K_i MMP7 = 60 nM K_i MMP8 = 3 nM K_i MMP9 = 3 nM K_i MMP12 = 5 nM K_i MMP13 = 7 nM K_i MMP14 = 10 nM
CGS 27023 A (MMI270B)	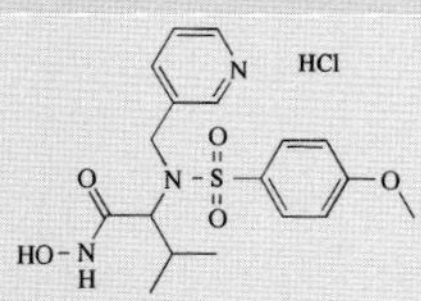	K_i MMP1 = 33 nM K_i MMP2 = 20 nM K_i MMP3 = 43 nM K_i MMP8 = 8 nM IC_{50} MMP9 = 8 nM IC_{50} MMP13 = 6 nM
COL 3 (metastat; CMT3)	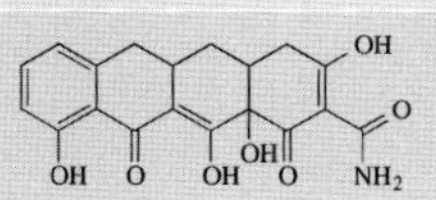	IC_{50} MMP1 = 34 μg per ml IC_{50} MMP8 = 48 μg per ml IC_{50} MMP13 = 0.3 μg per ml
Doxycycline	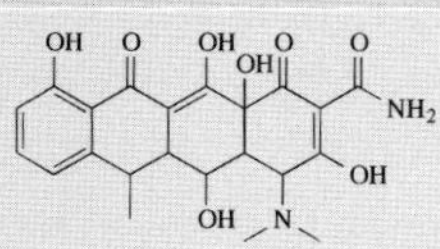	IC_{50} MMP1 > 400 μM IC_{50} MMP2 = 56 μM IC_{50} MMP3 = 32 μM IC_{50} MMP7 = 28 μM IC_{50} MMP8 = 26–50 μM IC_{50} MMP13 = 2–50 μM
RN 439 (*p*-aminobenzoyl-Gly-Pro-D-Leu-D-Ala-NHOH, MMP-Inh-1)	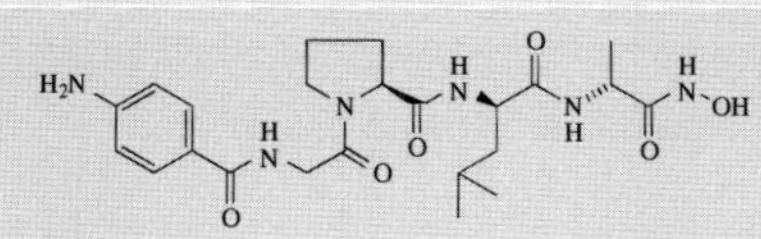	IC_{50} MMP1 = 1 μM IC_{50} MMP3 = 150 μM IC_{50} MMP8 = 1 μM IC_{50} MMP9 = 30 μM
GM6001 (ilomastat)	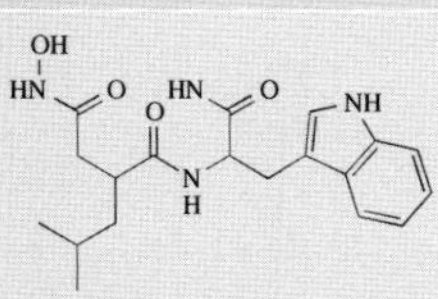	IC_{50} MMP1 = 0.4 nM IC_{50} MMP2 = 0.4 nM K_i MMP3 = 27 nM K_i MMP8 = 0.1 nM K_i MMP9 = 0.2 nM IC_{50} MMP14 = 5.2 nM
Marimastat (BB-2516; $C_{15}H_{29}N_3O_5$)	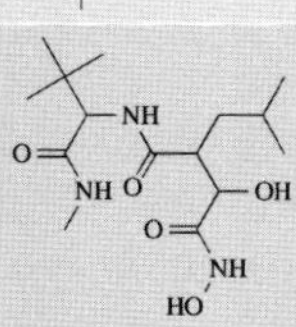	IC_{50} MMP1 = 5 nM IC_{50} MMP2 = 6 nM IC_{50} MMP3 = 200 nM IC_{50} MMP7 = 20 nM IC_{50} MMP8 = 2 nM IC_{50} MMP9 = 3 nM IC_{50} MMP14 = 1.8 nM
Minocycline	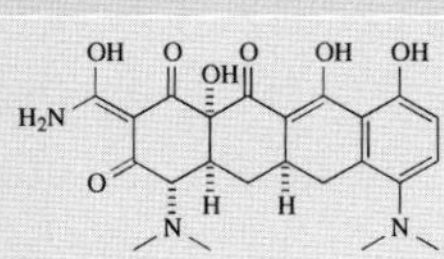	IC_{50} MMP1 >> IC_{50} MMP2 >> IC_{50} MMP3 = 290 μM IC_{50} MMP7 = 125 μM IC_{50} MMP9 = 180 μM

Figure 9.4. Structure and MMP inhibition of selected synthetic MMP inhibitors. (From Vandenbroucke and Libert, 2014.)

Inhibitor (alternative names)	Structure	MMP inhibition
MMP9 inhibitor I (CTK8G1150; $C_{27}H_{33}N_3O_5S$)	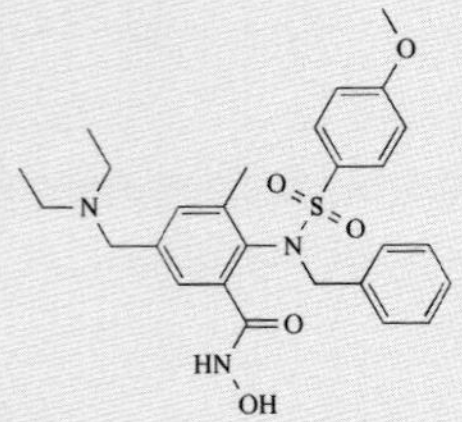	IC_{50} MMP1 = 1,05nM IC_{50} MMP9 = 5nM IC_{50} MMP13 = 113nM
ONO-4817 ($C_{22}H_{28}N_2O_6$)	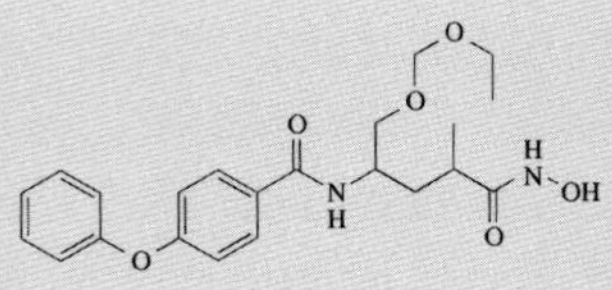	IC_{50} MMP1 = 1,600nM K_i MMP2 = 0.73nM K_i MMP3 = 42nM K_i MMP7 = 2,500nM K_i MMP8 = 1.1nM IC_{50} MMP9 = 2.1nM K_i MMP12 = 0.45nM K_i MMP13 = 1.1nM
Ro28 2653	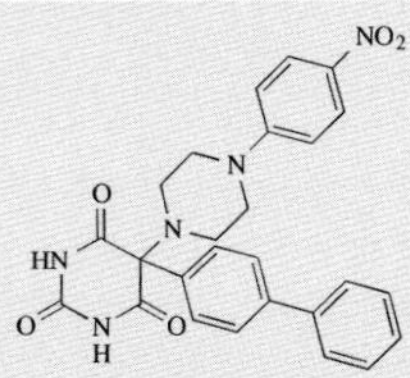	IC_{50} MMP2 = 7–246nM IC_{50} MMP8 = 15nM IC_{50} MMP9 = 12–23nM IC_{50} MMP14 = 96nM IC_{50} MMP16 = 91nM
SB 3CT	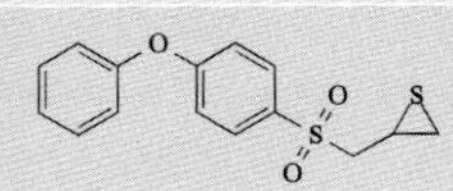	K_i MMP1 = 206μM K_i MMP2 = 14nM K_i MMP3 = 15μM K_i MMP7 = 96μM K_i MMP9 = 600nM

CMT, chemically modified tetracycline; IC_{50}, half-maximal inhibitory concentration; Ki, inhibition constant; MMP, matrix metalloproteinase.

Figure 9.4. (*Continued*)

hydroxamate zinc-binding group. One of these broad-spectrum inhibitors, MMI-270, is water soluble and orally available,[2] while others, such as MMI-166, was selective for MMP-2, MMP-9, and MMP-14 (Fig. 9.4).[2] Still other hydroxamate MMPIs (e.g., ABT-770, PD-166793 and prinomastat) were designed not to bind to the "shallow pocket" of MMP-1, with the reasoning being that by not inhibiting MMP-1 the musculoskeletal toxicity would be prevented. Another compound, Cipemastat (Ro 32–3555), which preferentially inhibited MMP-1, MMP-3, and MMP-9, was targeted for treating rheumatoid and osteoarthritis. However, its clinical trial was suspended because it did not stop/reduce progressive joint damage in patients with these diseases.

Despite the chronic problem of rapid drug metabolism of hydroxamate inhibitors, interest remained high because of their strong binding affinity for MMP active sites.[2,5] New hydroxamate inhibitors are being generated based on the analyses of structure–activity relationships (SARs) and quantitative SAR (QSARs), which predict the efficacy of new compounds based on knowledge of these molecular structures.[2,5,6] Whether or not they are actually therapeutically effective remains to be seen.

Other compounds that can bind zinc and stabilize it, thus avoiding the short half-life, appeared. These included carboxylates, hydrocarboxylates, sulphydryls, phosphoric acid derivatives, and hydantoins.[2,5,6] Further, when crystallography revealed the structures of MMPs, new inhibitors were no longer constrained by the need for peptidomimic backbones. Rebimastat (BMS-275291) was one of the first nonhydroxamate MMPIs (Fig. 9.4).[2] This MMPI has a thiol zinc-binding group, and is a broad-spectrum inhibitor. Of note, the structural scaffold of the thiol is a deep-pocket-binding, nonpeptide mimetic, which was designed not to block sheddase activities, and thus to allow MMPs to still release cytokines and their receptors, all of which had important biologic activities.[2] Although this seemed like an exciting and fail-safe concept, Phase II trials in early-stage breast cancer and a Phase III trial in non-small-cell lung carcinoma resulted in toxicities. Another orally available pyrimidine-based inhibitor Ro 28–2653 blocked a host of MMPs, but spared MMP-1. Once again, despite promising results from preclinical studies, this compound failed to advance to a clinical trial. Still other MMPIs used an alternative zinc-binding group (hydantoin or biphenyl sulphonamide carboxylate), which was selective for MMP-13 and which was targeted as therapeutic agents for osteoarthritis.[2] But these, too, failed in trials. All in all, inhibitors based on zinc binding, although potentially potent, simply did not work.

Another approach that emerged in the mid-2000s was phosphorous-based MMPIs.[2,5,6] These are structural mimics of the tetrahedral transition state of amide hydrolysis, where the phosphinic moiety binds the zinc ion (Fig. 9.5).[2,5,6] By modifying the R1 and R2 substituents, potency at nanomolar concentrations could be achieved,

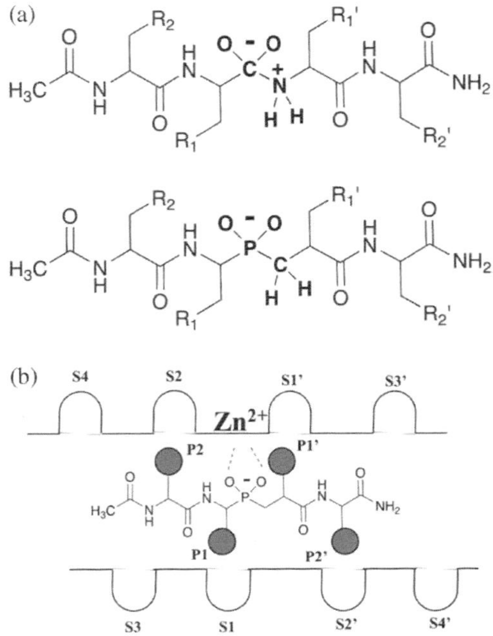

Figure 9.5. Phosphinic Bond. (a) Similarity between the structure of a peptidic sequence in the so-called tetrahedral-intermediate (above) and a phosphinic peptide (bottom); (b) schematic representation between the subsites of a zinc protease and a phosphinic peptide sequence. (From Yiotakis and Dive, 2009 with permission.)

but even in preclinical trials, any efficacy depended on careful timing of dose.

The tetracycline antibiotics, minocycline and doxycycline, have some ability to inhibit MMP enzymatic activity.[1,2,9] Importantly, even after all these years and numerous clinical trials, doxycycline hyclate is the only MMPI approved by the US Food and Drug Administration for any human disease, and it is indicated for treatment of periodontal disease.[1,2,5,8,9] Consequently, more recent efforts have turned to alternative strategies that focus on less conserved MMP subsites.[2,3,5,10] In addition, X-ray crystallography, along with NMR and computational methods, have facilitated modeling of drug–protein interactions with MMPIs that bind to noncatalytic sites.[2,5,10] These compounds, with as much as three orders of magnitude greater potency and selectivity, have been designed, and several compounds bind to exosites, that is, sites located outside the substrate-binding pocket.[2,4,6,7,10,11]

9.3 Novel Approaches to Designing MMPIs

As noted, the overall similarity of the active site structure among the MMPs has prevented selective inhibition of a particular MMP. One alternative has been to identify exosites that influence the function and/or substrate specificity of particular MMPs. Thus, developing antibodies that target allosteric/exosites sites has been one approach that holds substantial promise.[4] "Allostery" is defined as sites, alternative to the enzymatically active site, that are present on the surface of the enzyme and that help confer regulation of enzyme function. Although these sites are found away from the catalytically active site, they are critically important because they affect enzyme activity by changing protein conformation.[4] The molecular structure and organization of MMPs into modular domains (Fig. 9.2) allow these enzymes to use allosteric interactions to help mediate their activities. As one example, the long and flexible O-glycosylated domain of MMP-9 facilitates protein–substrate interactions by allowing the terminal regions of the enzyme to move, and thus to enhance the ability of the enzyme to bind to collagen fibrils and promote the hydrolysis of collagen fibers.[4] Indeed, flexibility between the hemopexin domain and the catalytic domain during catalysis may be important in facilitating the degradation of collagen by the classical interstitial collagenases (MMP-1, MMP-8, and MMP-13). Another example of allosteric influence is the activation of proMMP-2. ProMMP-2 activation on cells that are expressing MMP-2 and MT1-MMP can be induced by a monoclonal antibody directed at the hemopexin domain of MT1-MMP.[4] The antibody stabilizes MT1-MMP dimers to increase the efficiency of activating proMMP-2.

Another mechanism for regulating the activation of proMMPs and their enzymatic activity is homodimerization/multimerization, especially for MT1-MMP and MMP-9.[4] The hemopexin, transmembrane, and cytoplasmic domains are necessary for MT1-MMP to form a homo-oligomer complex, which then facilitates cell surface activation of pro-MMP-2 by MT1-MMP, and subsequent collagen degradation. These homodimerizations and multimerizations may also provide additional target sites for interfering with allosteric interactions. Recently a few agents with potential for success in specificity and relatively little toxicity have emerged. One example is a small molecule that targets the hemopexin domain of MMP-9. It does not modulate enzyme activity, but it blocks MMP-9 homodimerization, and thus blocks downstream signaling pathways by which MMP-9 mediates cell migration.

Peptide substrates with a triple-helical structure have conformational features that interact with exosites on MMPs, and these substrates help to identify allosteric inhibitors. Indeed, an inhibitor that binds to exosites on MMP-13 showed selectivity for this enzyme.[4] This inhibitor was more effective against MMP-13 triple-helical peptidase activity compared with MMP-13 single-stranded peptidase activity, suggesting that it may interact with a collagen-binding site on MMP-13.

Thus, focusing on exosites as therapeutic targets in MMPs may provide the requisite specificity that has been so absent in realizing the potential of MMPs as successful therapeutic targets. Antibodies are well known for their therapeutic efficacy in many different diseases, and extending this approach to developing blocking antibodies that target particular regulatory sites on MMPs is logical (Fig. 9.6).[4] Theoretically, antibodies could block proteolytic activity by binding at or near the catalytic site, preventing access to substrate. Alternatively, antibodies could block enzyme activity indirectly by binding to sites that could allosterically interfere with enzyme activity, that is, exosites, such as the hemopexin domain, the fibronectin domain, or the linker region, in order to prevent substrate-specific binding.

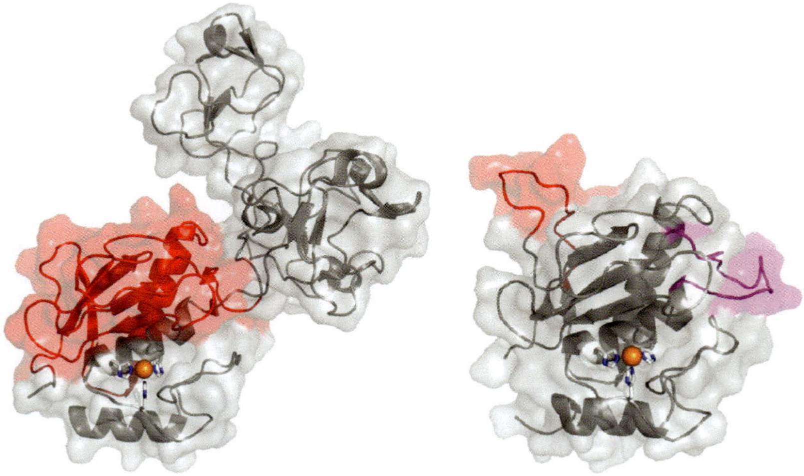

(a) (b)

Figure 9.6. Target allosteric protein sites used for the generation of anti-MMP function blocking Abs. A. Interaction site of anti-MMP-9 mAb REGA-3G12 as depicted in the structure of the catalytic and fibronectin domains of MMP-9. The site of interaction as analyzed by biochemical analysis is shown in red. Importantly, the antibody interacts with the catalytic domain surface, but not with the zinc active site (zinc ion represented as orange sphere ligated by three conserved histidines shown as sticks). B. The catalytic domain of MT1-MMP is shown in cartoon and surface representation (grey), the catalytic zinc ion is shown as orange sphere ligated by three conserved histidines (shown as sticks). The two exposed loops that were used to generate anti-MT1-MMP antibodies are colored in red and magenta respectively. These surface loops reside away from the catalytic zinc site. (From Sela-Passwell *et al.*, 2010.)

The most selective inhibitor of human MMP-9 that has been developed is the high-affinity monoclonal antibody REGA-3G12. It blocks the biological activity of MMP-9 but does not affect MMP-2. The antibody targets the amino-terminal region of the catalytic domain of MMP-9 but, interestingly, does not include the zinc ion-binding region.[4] Other research has developed a monoclonal antibody against MT1-MMP directed at loops in the catalytic domain of the enzyme.[4] This antibody blocks proteolytic activity and impairs angiogenesis and cell invasion.

Still another antibody, DX-2400, directed at the catalytic domain of MT1-MMP has been developed by Dyax Corporation.[4] Impressively, the antibody competitively inhibited this enzyme with Ki in the subnanomolar range (Ki = 0.8 nmol/L) by directly blocking the catalytic site but of note, it did not interfere with the activity of several other secreted or membrane-bound human MMPs. Functionally, DX-2400 prevented the ability of MT1-MMP to activate pro-MMP-2, inhibited angiogenesis, and slowed metastasis in animal models.[4,12] Thus, this antibody represents a potentially important therapeutic advance.

9.4 Strategies for Blocking MMP Gene Expression

At about the same time that interest in MMPIs was exploding, attention was also focused on blocking MMP gene expression.[1,3,13–16] Although this is an attractive concept since it blocks the actual synthesis of an MMP, the same issues of specificity and selectively that plagued blocking enzyme activity apply to blocking gene expression. Nonetheless, targeting extracellular factors and signal-transduction pathways that are upstream of MMPs and/or the nuclear factors that activate transcription has become feasible (Figs. 9.7 and 9.8).

The early efforts at blocking gene expression focused on the promoter regions of MMPs. Glucocorticoids were the initial drug of choice and they effectively suppressed MMP gene expression, largely hindering the binding of AP-1 proteins (Fos and Jun family members) to the proximal AP-1 site. However, many different extracellular stimuli (growth factors and cytokines), and signaling pathways that activate MMPs also activate the proximal promoter of many genes and therefore, have several side-effects resulting from the lack of specificity.

Simultaneously, there was growing interest in the use of retinoids, Vitamin A-like compounds that bind to nuclear receptors RARs, and RXRs. While the exact mechanism of suppressing MMP expression was uncertain, retinoids were effective in blocking transcription. However, although retinoids reduced inflammation and joint destruction in several animal models of arthritis, disturbing side-effects, such

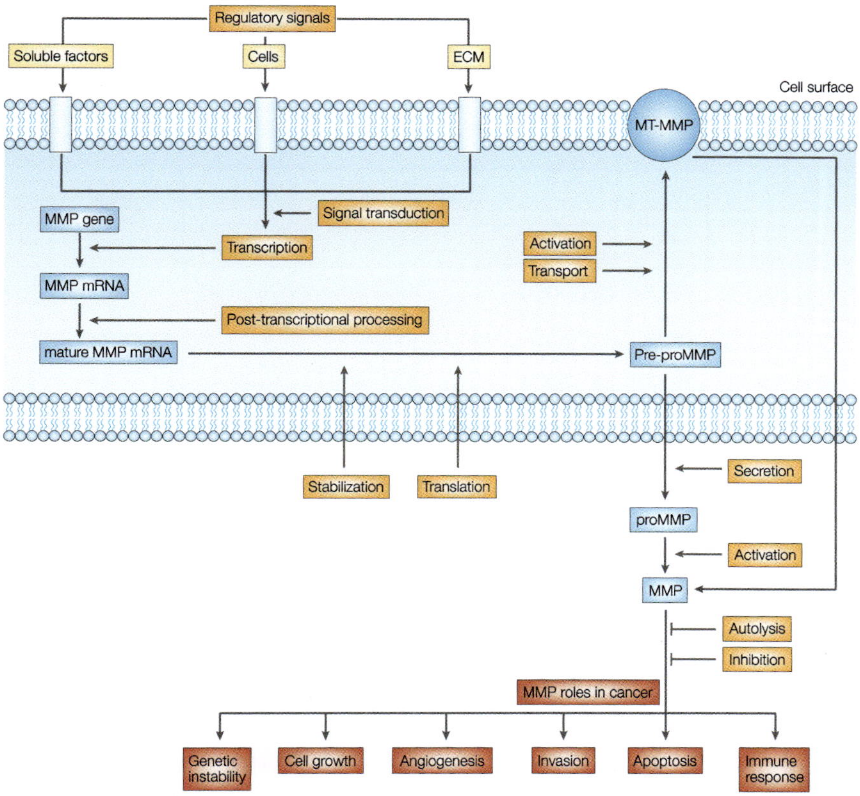

Figure 9.7. Levels of regulation of MMP expression and activity. Diverse regulatory signals, such as soluble factors, extracellular-matrix (ECM)–cell interactions or cell–cell contacts (cells) interact with specific receptors at the cell surface and initiate a cascade of events that lead to the generation of functional MMPs, which are localized to the cell surface (MT-MMPs) or secreted to the extracellular medium (proMMP). ProMMPs are activated by different events. These active MMPs are involved in a number of processes that promote cancer development (red boxes), including promoting genetic instability, cell growth, angiogenesis, and invasion. They also interfere with apoptosis induction and the host antitumor immune response. MMP autolysis or inhibitors can interfere with the induction of these cellular effects. The levels of MMP regulation that might be therapeutically targeted are shown in orange boxes, and include cell responses to regulatory signals, signal transduction, transcription induction, post-transcriptional processing, MMP activation and transport, and secretion. (From Overall and López-Otín, 2002.)

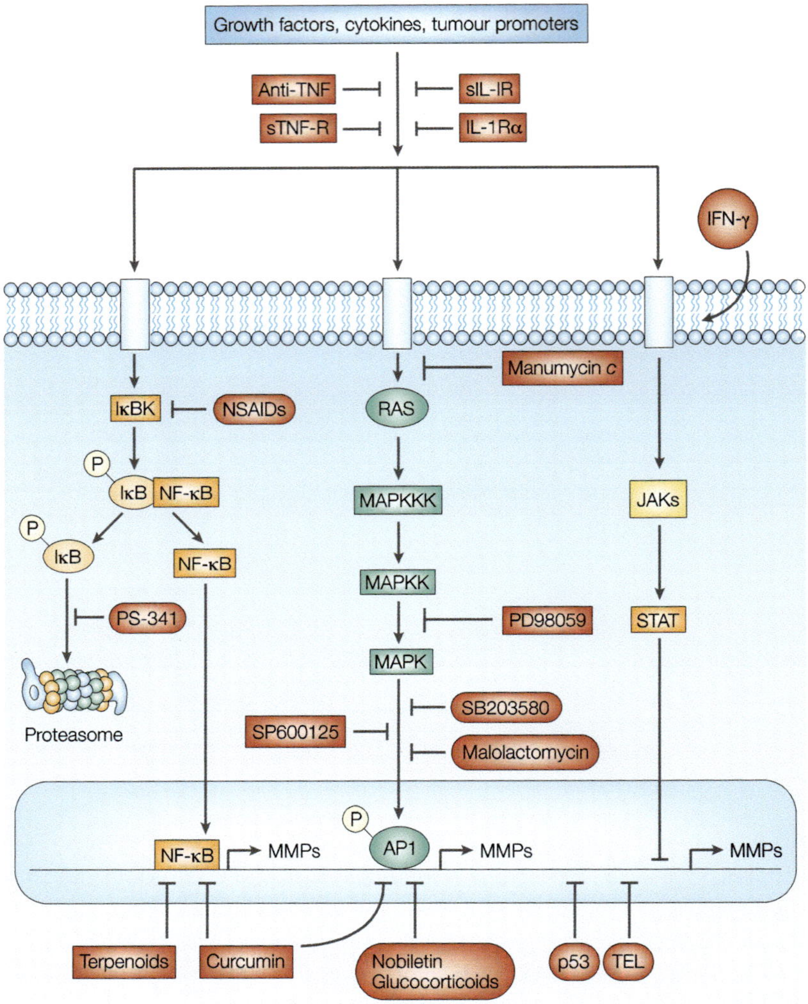

Figure 9.8. Signaling pathways involved in MMP gene transcription, and potential strategies for therapeutic intervention. Compounds that are able to block the transcription of matrix metalloproteinase (MMP) genes at different levels are shown in red boxes. Extracellular factors, such as interferon-γ (IFN-γ) inhibit MMP transcription via the JAK–STAT signaling pathway. Monoclonal antibodies against tumor necrosis factor-α (anti-TNF), soluble forms of the TNF receptor (sTNF-R), natural antagonists of the interleukin (IL)-1 receptor-α (IL-1Rα) or soluble forms of this receptor (sIL-R) can block signaling pathways initiated by extracellular factors such as TNF-α and IL-1, which induce MMPs in cancer cells. Compounds such as

as bony outgrowths of the spine and abnormal lipid profiles dampened enthusiasm.[14] Consequently, the promiscuous nature of RARs and RXRs and the pleiotropic effects of glucocorticoids rendered them out of favor as therapeutic agents. Directly blocking MMP gene expression was abandoned as a therapeutic strategy.

An emerging alternative approach is blocking gene expression by blocking the signal transduction pathways stimulated by cytokines and growth factors, which lead to down-stream induction of MMP gene expression (Figs. 9.7 and 9.8).[3] The "biologic" therapies that show efficacy in treating autoimmune/inflammatory diseases, such as rheumatoid arthritis, psoriasis and Crohn's Disease are one example of this success (Table 9.2).[17–19] These therapies are based on monoclonal antibodies that target TNFα, soluble forms of TNF, and/or of the IL-1β receptor. Because the antibodies block expression of other genes in addition to MMPs, they are not specific. However, since MMPs are a downstream target of these signaling pathways, joint destruction is lessened. On the other hand, they suppress inflammatory and immune responses in general, and although effective, they can be associated with serious side effects, such an increased incidence of tuberculosis, lymphoma, and leukemia, and even some neurological disorders.

Figure 9.8. (*Continued*) manumycin A, SB203580, malolactomycin, SP600125 or PD98059 act at different levels to block the signal-transduction pathways that are associated with MMP transcriptional induction in human tumors. Finally, there are several possibilities to target the nuclear factors that are responsible for MMP transcriptional upregulation. Glucocorticoids, terpenoids, curcumin, nobiletin, or NSAIDs (nonsteroidal anti-inflammatory drugs) block the activity of transcription factors such as AP1 and NF-κB, which regulate the transcription of several MMP genes. Similarly, restoring the activity of transcription factors such as p53 and TEL, which negatively regulate MMP expression and the activity of which is lost in human tumors, could downregulate these genes. IFN-γ, interferon-γ; IκB, inhibitor of κB ; IκBK; inhibitor of κB kinase; JAK, JUN-activated kinase; MAPK, mitogen-activated protein kinase; MAPKK, mitogen-activated protein kinase kinase; MAPKKK, mitogen-activated protein kinase kinase kinase; NF-κB, nuclear factor of κB; STAT, signal transducer and activator of transcription; TEL, translocation-ETS-leukemia. (From Overall and López-Otín, 2002.)

Table 9.2. Overview of the currently available biologic DMARDs for the treatment of rheumatoid arthritis.

Name	Target	Format	Mechanism	Administration	Approximate half-life*
Etanercept	TNF	Recombinant human fusion protein of the TNF receptor and the Fc portion of IgGI.	Works as a decoy receptor. It binds to sluble TNF, blocking the binding to its receptor	sc. Injection once (50 mg), or twice (25 mg) a week	3–6 days
Adalimumab	TNF	Fully human IgGI MAb	Binding to TNF	sc. Injection once every second week	13 days
Infliximab	TNF	Chimeric murine-human IgGI MAb	Binding to soluble and mb bound TNF	iv. infusion every 4–8 weeks	9 days
Golimumab	TNF	Fully human IgGI MAb	Binding to soluble and mb bound TNF	sc. Injection once a month or iv. Infusion at 0 and 4 weeks, thereafter every 8 weeks	13 days
Certolizumab-pegol	TNF	Humanized pegylated anti-TNF Fab' fragment	Binding to TNF	sc. Injection once every second week	14 days
Anakinra	IL-I	Recombinant human IL-I receptor antagonist	Binding to IL-I type-I receptor	sc. Injection once a day	4–6 hours

Tocilizumab	IL-6	Humanized recombinant IgGI MAb	Binding to soluble and membrane bound IL-6 receptor	iv. infusion every 4 weeks	10–13 days
Rituximab	B cells	Chimeric murine-human IgGI MAb	Binding to CD20 and depletion of CD20+ B cells	Two initial infusions 14 days apart. Courses may be repeated every 6 months or more frequently depending on disease activity	18 days (range: 5–76 days)
Abatacept	T cells	Recombinant human fusion protein of the extracellular domain of CTLA-4 and the Fc portion of IgGI	Binding to CD80/CD86, blocking T-cell co-stimulation	iv. infusion every 4 weeks or sc. Injection once a week	13 days (range: 8–25 days)

CTLA-4, cytotoxic T lymphocyte associated antigen 4; DMARDs, disease-modifying anti-rheumatic drugs; IgG, immunoglobulin G; IL, interleukin; iv, intravenous; MAb, monoclonal antibody; mb, membrane; sc, subcutaneous; TNF, tumor necrosis factor

*The half-lives provided here refer to the biological effect and the physical half-life reported by the manufacturers which may differ depending on the format of the drug.

Source: From Vivar and Van Vollenhoven (2014).

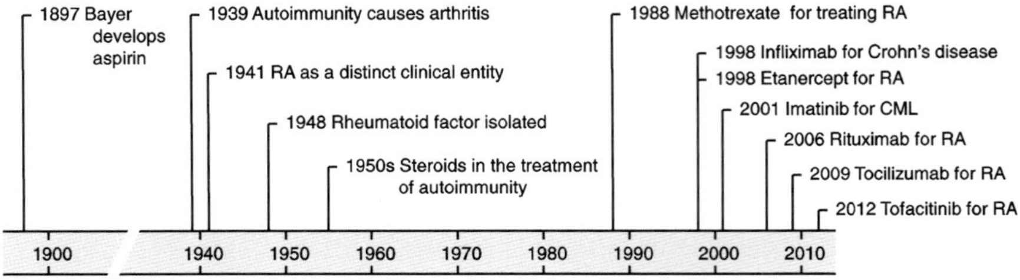

Figure 9.9. Timeline of rheumatoid arthritis research and therapy. RA, rheumatoid arthritis; CML, chronic myelogenous leukemia. (From Mócsai *et al.*, 2014.)

Small-molecule inhibitors that target a specific mutation, particularly in certain cancers, are dramatically effective as therapeutics. They provide an important proof-of-principle, even though information about whether MMP is an indirect target of these therapies may not be known. Among the first (and perhaps among the most famous) is the signal transduction inhibitor, STI571 (also called Gleevec or Imatinib).[20] This compound is a selective inhibitor for *abl* tyrosine kinases, including *bcr-abl*, as well c-*kit* and the platelet-derived growth factor receptor tyrosine kinases, and was initially used to treat patients with chronic myelogenous leukemia (CML), with minimal toxicities. The fusion protein of *bcr-abl*, the product of a chromosomal translocation, is a constitutively active tyrosine that binds ATP and transfers a phosphate from ATP to kinase residues on various substrates. As a result, there is increased proliferation of myeloid cells characteristic of CML. STI571 blocks the binding of ATP, thereby blocking the activity of the *bcr-abl* kinase. Without this kinase activity, the downstream substrates cannot be phosphorylated and subsequent cell proliferation is stopped. Not surprisingly, resistance often arises and even a small amount of residual tumor continues to mutate. Interestingly, the compound has also been used to treat rheumatoid arthritis (Fig. 9.9).[19]

9.5 Conclusions

The relative nonspecific nature of the signaling pathways that induce MMP expression illustrate some of the difficulties associated with suppressing MMP gene expression. However, definitive success with "biologics" that block signaling pathways of inflammatory cytokines is apparent, as are emerging reports on the development of monoclonal antibodies that are directed at exosites and that block enzymatic activity. Another important realization is that not all MMPs need to be blocked in all cases and at all times.[2,21] Indeed, depending on their substrates, MMPs can have opposing effects on central aspects of tumor progression, such as tumor growth and survival, angiogenesis, invasion, and modulation of the immune response, and in some cases, MMPs are actually "antitargets".[21] Thus, it appears to be more essential than ever to design MMPIs that are specific for a particular enzyme. The approach of blocking enzyme activity rather than gene expression seems preferable, since this favors the ability to block only a particular enzyme. The recent studies on compounds, be they small molecule inhibitors or targeted monoclonal antibodies, bode well for eventual success, much as the biologics have revolutionized the treatment of autoimmune diseases. Nonetheless, even if a particular MMP can be inhibited, issues of its expression throughout the body and of possible deleterious effects of blocking potential beneficial activities will always need to be considered.

References

1. Brinckerhoff, CE and LM Matrisian (2002). Matrix metalloproteinases: a tail of a frog that became a prince. *Nature Reviews Molecular Cell Biology,* 3, 207–214.
2. Vandenbroucke, RE and C Libert (2014). Is there new hope for therapeutic matrix metalloproteinase inhibition? *Nature Reviews Drug Discovery,* 13, 904–927.
3. Overall, CM and C López-Otín (2002). Strategies for MMP inhibition in cancer: innovations for the post-trial era. *Nature Reviews Cancer,* 2, 657–672.
4. Sela-Passwell, N, G Rosenblum, T Shoham and I Sagi (2010). Structural and functional bases for allosteric control of MMP activities: can it pave the path for selective inhibition? *Biochimica Biophysica Acta — Molecular Cell Research,* 1803, 29–38.

5. Yiotakis, A and V Dive (2009). Synthetic active site-directed inhibitors of metzincins: achievement and perspectives. *Molecular Aspects Medicine,* 29, 329–338.
6. Tallant, C, A Marrero and FX Gomis-Rüth (2010). Matrix metalloproteinases: fold and function of their catalytic domains. *Biochimica Biophysica. Acta — Molecular Cell Research,* 1803, 20–28.
7. Rodríguez, D, CJ Morrison and CM Overall (2010). Matrix metalloproteinases: what do they not do? New substrates and biological roles identified by murine models and proteomics. *Biochimica Biophysica Acta — Molecular Cell Research,* 1803, 39–54.
8. Murphy, G and H Nagase (2008). Progress in matrix metalloproteinase research. Molecular Aspects Medicine, 29, 290–308.
9. Griffin, MO, E Fricovsky, G Ceballos and F Villarreal (2010). Tetracyclines: a pleitropic family of compounds with promising therapeutic properties. Review of the literature. American Journal of Physiology Cell Physiology, 299, C539–C548.
10. Murphy, G (2010). Fell-Muir Lecture: Metalloproteinases: from demolition squad to master regulators. *International Journal of Experimental Pathology* 91, 303–313.
11. Overall, CM (2002). Molecular determinants of metalloproteinase substrate specificity: matrix metalloproteinase substrate binding domains, modules, and exosites. *Molecular Biotechnology,* 22, 51–86.
12. Nishihara, T *et al.* (2015). Matrix metalloproteinase-14 both sheds cell surface neuronal glial antigen 2 (NG2) proteoglycan on macrophages and governs the response to peripheral nerve injury. *Journal of Biological Chemistry,* 290, 3693–3707.
13. Vincenti MP, LA White, DJ Schroen, U Benbow and CE Brinckerhoff (1996). Regulating expression of the gene for matrix metalloproteinase-1 (collagenase): mechanisms that control enzyme activity, transcription, and mRNA stability. *Critical Reviews in Eukaryotic Gene Expression*, 6, 391–411.
14. Brinckerhoff, CE and MB Sporn (2003). Retinoids and rexinoids for the 21st century: a brave new world for arthritis. *Journal of Rheumatology*, 30, 4–7.
15. Boyd, DD (2006). Regulation of matrix metalloproteinase gene. *Journal of cell Physiology*, 19–26. doi:10.1002/JCP
16. Schroen DJ and CE Brinckerhoff (1996). Nuclear hormone receptors inhibit matrix metalloproteinase (MMP) gene expression through diverse mechanisms. *Gene Expression*, 6,197–207.17.

17. Wang, DS, Y Li, Y Liu and GX Shi (2014). The use of biologic therapies in the treatment of rheumatoid arthritis. *Current Pharmaceutical Biotechnology,* 15, 542–548.
18. Vivar, N and RF Van Vollenhoven (2014). Advances in the treatment of rheumatoid arthritis. *F1000Prime Reports,* 6, 31.
19. Mócsai, A, L Kovács and P Gergely (2014). What is the future of targeted therapy in rheumatology: biologics or small molecules? *BMC Medicine,* 12, 43.
20. Mauro, MJ and BJ Druker (2001). STI571: targeting BCR-ABL as therapy for CML. *Oncologist,* 6, 233–238.
21. Dufour, A and CM Overall (2013). Missing the target: matrix metalloproteinase antitargets in inflammation and cancer. *Trends Pharmacological Science,* 34, 233–242.

10. Through the Looking Glass into the Era of Precision Medicine

Constance E. Brinckerhoff

10.1 Introduction

The number of matrix metalloproteinases (MMPs), along with their myriad of substrates and ever-changing levels of expression and activity, attest to the difficulty in trying to assess their future roles in biology and medicine. This is a complex issue that will need continuous evaluation in order to place MMPs, at any given time, within their appropriate context of normal physiology and disease pathology. Even a brief summary of our current understanding of "MMP-ology" underscores this complexity.

More than 20 distinct MMPs have been identified, all part of the metazincins, a larger family of metalloenzymes in which Zn is an integral part of their structure and which requires Ca^{++} for activity. Thus, MMPs are a substantial and important group of proteolytic enzymes. MMPs were originally defined by their ability to degrade components of the extracellular matrix (ECM). However, a myriad of substrates has been identified as targets of MMPs, many of which are not related to degradation of the (ECM), and are, therefore, independent of and/or in conjunction with their proteolytic activities. These include diverse roles as transcription factors, activators of signal-transduction pathways, and even as signaling molecules, and this functional diversity further complicates efforts to regulate their expression and/or activities for therapeutic benefit. Indeed, the levels of gene expression and enzymatic activities of any given MMP vary considerably in the normal rhythm of physiology and in the various stages of disease pathology. Thus, figuring out where and when each MMP may "help" or "hinder" is a challenge. Furthermore, some MMPs have distinctly beneficial effects and therefore, are anti-targets in terms of therapeutic agents directed at blocking their effects. This is a point that is becoming increasingly recognized and cannot be overlooked or undervalued.

Nonetheless, rampant overexpression of MMPs is still linked incontrovertibly with disease and poor outcome, and consequently, the specific targeting of MMPs remains an important but largely unrealized goal. Results of early investigations with supposedly targeted inhibitors of gene expression were abandoned due to (unexpected) detrimental side effects on many other unrelated genes. Similarly, given the shared structural similarities among the catalytic site/substrate binding pocket, targeted blocking of a specific MMP was also dismally disappointing, leading to abandonment of ongoing research and clinical trials with inhibitors of MMP enzyme activities.

However, toward the end of the 20th century, biologic therapies began to appear, especially for the treatment of inflammatory/autoimmune diseases, such as rheumatoid arthritis. Originally designed to directly interfere with the effects of inflammation due to cytokines such as interleukin-1 (IL-1) and tumour necrosis factor-alpha (TNF-α), these "biologics" block signaling through receptors for these cytokines. Expression of MMPs was blocked or reduced since they are downstream targets of signaling through receptors for these cytokines, and the therapeutically beneficial effects in preventing or reducing joint destruction are well documented. Because of the commonality of shared signaling pathways, small-molecule inhibitors of signal/transduction pathways may also block MMP expression and thus, are beneficial. Therefore, whether "by hook or by crook", therapeutically blocking MMPs has become a reality, sometimes with a decidedly "good" clinical outcome. How do we take this mélange of existing data on MMP functions and some apparent therapeutic successes and apply this knowledge so that MMPs are positive partners in our biology? While this goal may not seem realistic, several new avenues of investigation may provide clues as to how to proceed.

10.2 Neurobiology: A New Frontier and Paradigm for MMPs

Certainly, surprises about MMPs remain to be discovered, and one emerging area of study is the potentially exciting functions of MMPs in neuropathology. However, not surprisingly, some of these functions may be beneficial, while others are detrimental, thus suggesting that MMPs are "double-edged" swords (Fig. 10.1).[1]

Neurodegeneration encompasses many disorders of the central nervous system (CNS), but in all instances, there is a chronic and progressive loss of neuronal cells, eventually resulting in a decline of the CNS. Prior to this degeneration, there is a period of neuroinflammation, and MMPs are an important component of this initial inflammatory process.[1] In addition, although the possible causes and etiologies of neurodegenerative diseases, such as Alzheimer's disease (AD), Parkinson's disease (PD), amyotrophic lateral sclerosis (ALS), Huntington's disease (HD), and multiple sclerosis (MS), are still unknown, MMPs have central, but sometimes intricate roles, in their progression. More than a dozen MMPs are involved in neurodegenerative diseases, including MMP-2, MMP-3, and MMP-9. The mechanisms by which MMPs aggravate neurodegenerative diseases are only beginning to be understood. MMPs are part of a common pathway of pathological changes that disrupt normal CNS homeostasis and that result in the accumulation of inflammatory molecules or aggregated proteins and peptides. The end result is increased permeability of CNS barriers and ultimately, neuronal cell death.

On the other hand, MMPs may have therapeutic value in helping to ameliorate neurodegenerative diseases.[1,2] Specifically, MMP-2 and MMP-9 have been implicated in the physiological catabolism of Alzheimer's amyloid-β (Aβ). Despite their association with disruption of the blood–brain barrier and hemorrhagic transformations after ischemic stroke, MMP-2 and MMP-9 selectively cleaved Aβ40. This cleavage generated fragments truncated only at the C terminus, ending at positions 34, 30, and 16, consistent with the hypothesis

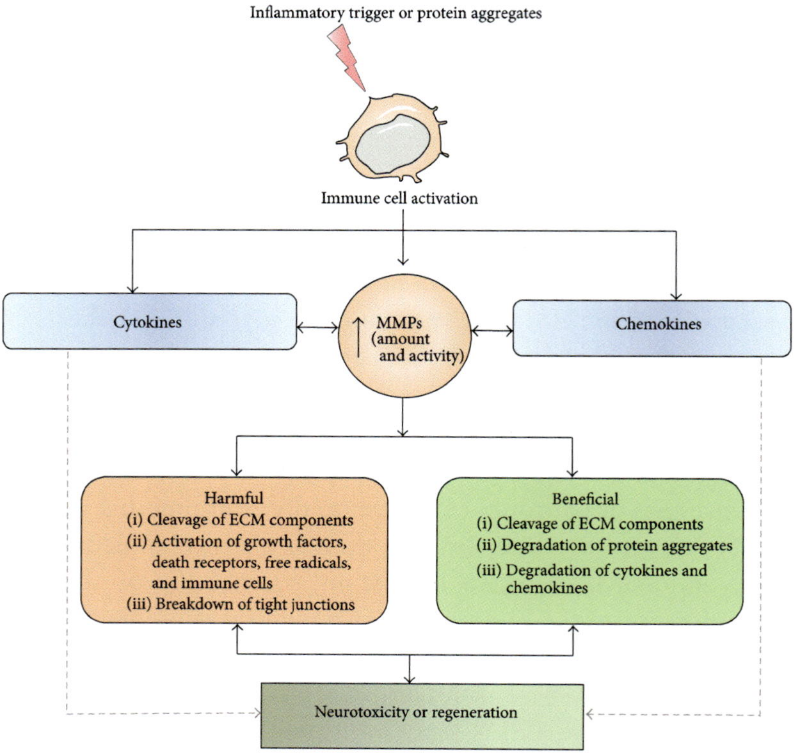

Figure 10.1. Schematic representation of the activation of MMPs, their interactions with cytokines and chemokines, and the outcome of the interactions. MMPs are induced and activated in the presence of an external trigger (e.g., inflammatory stimuli) or abnormal proteins (e.g., protein or peptide aggregates). The activated MMPs can alter the properties of cytokines and chemokines. They also interact with the extracellular matrix, cell surface receptors, growth factors, integrin, signaling molecules, and tight junction proteins and alter their properties. This affects neuroinflammation, cell death or survival, growth, and regeneration. ECM, extracellular matrix; MMP, matrix metalloproteinase; TJs, tight junctions. (From Brkic *et al.*, 2015.)

that C-terminally truncated Aβ fragments generated by MMPs were soluble. Further, these fragments did not display fibrillogenic properties or induce cytotoxicity in human cerebral microvascular endothelial or neuronal cells.[1,2] These intriguing findings support the concept that MMPs and truncated Aβ proteins may be associated with clearing mechanisms, rather than with fibrillogenesis. While these ideas may represent "pie-in-the-ski" science given the surprises revealed by past MMP activities, this reality no longer seems impossible.

There are also indications of new roles for MMP–tissue inhibitors of metalloproteinase (TIMP) interactions in conjunction with MMPs in neuropathologies. The traditional view has been that when the ratio between MMPs and TIMPs is disturbed in favor of more MMPs by pathological conditions, there is a net increase in proteolytic activity[3] and that this increase is linked to disease progression. However, an intriguing alternative view assigns important biological roles for TIMPs. These functions are independent of their ability to block the enzymatic activities of MMPs and involve the binding of

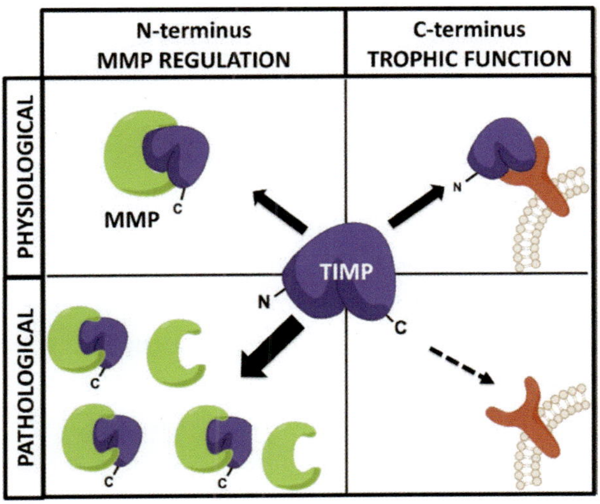

Figure 10.2. MMP regulation of TIMP-mediated signaling. Under physiological conditions, there is a homeostatic balance between the production of MMPs and expression of TIMPs. The regulation of MMP proteolysis by TIMPs is mediated through their N-terminus, whereas the C-terminus of TIMPs interacts with cell surface receptors to initiate intracellular signaling cascades. Under conditions of pathology, increased MMP expression results in a net increase in MMP-mediated proteolysis, which also results in the depletion of TIMP proteins available to provide receptor-mediated signaling. This loss of TIMP-mediated signaling may have important implications for disease pathology. (From Moore and Crocker, 2012.)

TIMPs to their cognate receptors (Fig. 10.2).[3] This view implies that it is the MMPs that can block the functions of TIMP, and that MMPs are inhibiting the signaling functions of TIMPs. During inflammatory demyelination, inhibiting MMPs is anti-inflammatory, presumably through TIMPs, but blocking MMPs with pharmacologic agents can also mitigate demyelination in experimental autoimmune encephalomyelitis. This has led to the speculation that part of this beneficial effect might be attributed to signaling by TIMPs that are not bound to MMPs, whose activites are now blocked by another mechanism. Indeed, several MMP-independent actions of TIMP-1 in the CNS have been documented.[3] This novel view reflects emerging data indicating that TIMP signaling and inhibiting MMPs represent two important, but distinct, functions of TIMPs, which may provide a new rationale for the use of TIMPs as therapeutic agents.

10.3 MMPs as Agents of Imaging and as Vehicles for Selective Drug-Delivery

Concomitant with these studies that ascribe new functions for MMPs and that hint at intriguing new therapeutic uses for MMPs, other ingenious and clever applications for MMPs are being developed. MMPs are beginning to be used as agents for enhancing *in vivo* imaging and as vehicles for delivery of drugs. In general, molecular imaging probes should have certain key characteristics: (1) rapid binding with high affinity; (2) specificity for the target; (3) rapid clearance of unbound molecules; (4) high target-to-background ratio; (5) high stability coupled with low immunogenicity and toxicity; and finally, (6) practicality in terms of production and cost. Probes can be made from small molecules, peptides, proteins, antibodies, the antigen-binding region of antibodies (Fab fragments), nanobodies, and nanoparticles (Fig. 10.3).[4] For optimal efficacy, MMPs will need to be incorporated into these modalities.

Since heightened expression and activity of MMPs often occur within the tissue microenvironments of diseases such as cancer and rheumatoid arthritis, it is not surprising that these aberrant levels of

Molecular imaging probes:	
Small molecules	Low molecular weight substances (≤900 Da) that are synthetised by chemical reactions between organic and/or inorganic compounds, e.g. methylene diphosphonate.
Peptides	Natural or synthetic compounds containing two or more amino acids, e.g. arginine-glycine-aspartic acid (RGD) tripeptide.
Proteins	Compound consisting of one or more chains of amino acids, e.g. annexin.
Antibodies	Proteins, immunoglobulins, that recognize and bind with high affinity to the antigen they were raised against. Antibodies consist of two heavy chains and two light chains (~150 kDa), e.g. anti-TNF-α antibodies.
Antibody fragments	Proteins comprising only one constant and one variable domain of the heavy and light chains of an antibody (~50 kDa), e.g. anti-E-selectin Fab fragments.
Nanobodies	The cloned V_{HH} domain of a heavy-chain antibody occuring in certain animal species (~15 kDa), e.g. MMR-targeting nanobodies.
Nanoparticles	Particles of 1 to 100 nm that possess unique material characteristics, e.g. chitosan nanoparticles.

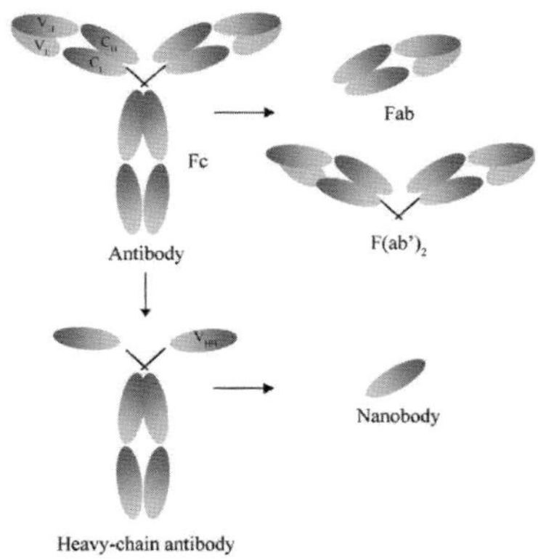

Figure 10.3. Composition of probes. Probes can be composed of small molecules, peptides, proteins, antibodies, antibody fragments, nanobodies, and nanoparticles. A schematic overview of a conventional antibody, a heavy-chain antibody, Fab fragments, and a nanobody is given. CH, heavy chain constant domain; CL, light chain constant domain; Fab, antigen-binding domain; Fc, constant domain; MMR, macrophage mannose receptor; TNF-α, tumor necrosis factor-alpha; VH, heavy chain variable domain; VHH, heavy chain only antibody VL, light chain variable domain. (From Put *et al.*, 2014.)

expression are being harnessed as agents for molecular imaging and assessing a therapeutic response.[4,5] To date, these novel applications of MMPs have been documented only in experimental animal models, but their extension to human disease may not lag far behind.

In arthritis, new molecular imaging techniques open up the possibility for visualizing early diagnosis and degradative matrix changes in rheumatoid arthritis (RA).[4] Earlier diagnoses through the use of these noninvasive techniques will permit continual monitoring of disease progression, provide guidance in the selection of treatment options, and may predict the clinical outcome resulting from a treatment option. In cancer, current imaging techniques have difficulty detecting tumors that are smaller than 1 cubic mm. Amplifying the signal of a probe could, therefore, help to identify occult tumors and metastases. One approach to signal amplification is activity-based probes (ABP) linked to MMPs, which take advantage of MMP substrates to enhance imaging.[5,6] Constructing MMP-ABPs involves labeling peptide substrates, which are either broad spectrum or MMP selective. The substrates are then cleaved by the MMPs, which are already present in the microenvironment. The peptides are labeled with fluorophore/quenchers, and their cleavage by MMPs gives a readout of enzymatic activity that is accompanied by a signal that is imaged.[5] For example, incorporating near-infrared fluorophores with wavelengths between 650 and 900 nm facilitated the imaging of MMP-7 activity *in vivo*, detecting tumors as small as 0.01 cm^2 in a mouse model of colon cancer. Activatable near-infrared broad-spectrum probes, which can detect MMP-2, -3, -7, -9, -12, and -13, are now commercially available for *in vivo* studies, emphasizing the translational nature of this work.[5]

Radiolabeled therapeutics is another imaging option that could provide early and more specific diagnoses and facilitate monitoring of treatment efficacy. The relative ease of synthesizing radiolabeled probes makes them particularly attractive in terms of the possibility of rapid translation to the clinic. This approach is based on images through single-photon emission computed tomography/positron emission tomography (SPECT/PET).[4,5] Labeling probes that are cleaved by MMPs with radioisotopes is attractive. First attempts with this approach used small-molecule MMP inhibitors conjugated to radioisotopes. Subsequent studies that used a noninhibitory MMP activatable probe designed labeled with ^{18}Fluoride (^{18}F) allowed monitoring of therapeutic response to protease inhibitors.[5] It also overcame limitations of inhibitor-based probes, which often bind to both active and latent forms of MMPs and lead to inaccurate read-out of MMP activity.

Designing drugs to selectively target specific cells within their microenvironment, while minimizing systemic toxicities remains the gold standard of therapies. Delivering drugs to a specific microenvironment *in vivo* by means of nanoparticles such as liposomes and micelles, which are loaded with a drug cargo, can increase the longevity of the drug in the circulation and also enhance cellular uptake. Even more attractive is the possibility of loading the particles with targeting antibodies and an MMP-cleavable prodrug. Once the MMP has carried out is proteolytic assignment of cleaving the prodrug, and becomes active at the target site, it may amplify efficacy. For example, in a xenograft tumor model, a doxorubicin prodrug could not be internalized by cells because of the addition of a peptide that was supposed to prevent cytotoxicity. However, once the peptide was cleaved by MMP-2, -9, or -14 in the tumor microenvironment, the mature doxorubicin was internalized by cells, resulting in a higher therapeutic index than that of doxorubicin, alone. This kind of combinatorial therapy and prognosis hold promise as a win–win strategy.

However, technical difficulties remain.[4] These include identifying which MMP is appropriate to convert the prodrug to a mature drug, and designing a peptide sequence that is specifically processed by that MMP *in vivo*. Further, as suggested earlier, other rate-limiting issues are (1) the safety of murine and chimeric monoclonal antibodies; (2) potential problems with immunogenicity, which are being addressed by humanizing the antibodies; (3) the safety/toxicity profile of the drug pharmaceutically; (4) lack of efficacy of certain drugs in certain patients; and (5) high costs associated with developing and producing some of these designer therapies.

Nonetheless, despite these limitations, experimental breakthroughs are beginning to occur.[5] For example, in a zebra fish model, introducing a modified cysteine residue near the active site of MMP-12 and MMP-14 resulted in enzymes that were still proteolytically active and that could be specifically targeted with a small-molecule probe through the cysteine. Importantly, the probe irreversibly inhibited the enzymes and also allowed imaging of their localization. All of these techniques, reagents, and investigations illustrate the power of imaging and the potential of adding specific therapies as cargo in the nanoparticles in order to directly target the desired microenvironment, with limited systemic side effects. Walking the delicate line that balances the beneficial versus detrimental effects of MMPs as a "double-edged sword"[1] promises to be an ongoing journey that will need constant reassessment.

10.4 Actualizing the Power of MMPs as Prognostic Indicators in the Era of Precision Medicine

In every living being, cells throughout the body are continually dying a "natural" death, and when they do, they shed fragments of DNA into the blood. Injury or illness in specific cells and tissues generates even more of this DNA in blood.[7] Several investigator teams are now developing methods to trace this DNA to the tissue from which it came with the goals of early stage detection and monitoring of disease progression.[7] This circulating cell-free DNA (ccfDNA) is a valuable source of material that is easily obtained from a simple blood sampling.[7–10] The ccfDNA released from cells that are dying is being used as a sensitive and noninvasive diagnostic tool for monitoring progression of a variety of cancers, graft failure following transplantation, and possible genetic abnormalities in pregnant women.

The experimental techniques use a quantitative polymerase chain reaction (Q-PCR) multimarker approach to qualitatively and quantitatively evaluate a series of parameters that may include measuring total ccfDNA concentration, mutant ccfDNA concentration, the amount of fragmented DNA, and the relaltive proportion of mutant ccfDNA. Specific markers of interest that have been successfully detected include the V600E mutation in cancer cells, levels of ccfDNA fragmentation, donor-specific genetic markers in transplant recipients, genomic alterations in fetal DNA, and identification of cell type-specific DNA methylation signatures (Fig. 10.4).[7] Since this technique can be used to identify ccfDNA derived from any cell type in the body, it provides an opportunity for diagnosis with minimal amounts of material, and may well lead to increased knowledge of the molecular mechanisms underlying many human diseases.

Imagine, then yet another level of medical and technical elegance superimposed on this elegant technology: the ability to accurately and sensitively measure levels of circulating MMPs and then to correlate these data to the very cells that are producing them and releasing their ccfDNA. Indeed, the prognostic value of correlating levels

Figure 10.4. Schematic of methyl groups on DNA. The methyl groups placed on DNA (yellow, green) by an enzyme (purple) differ by tissue and can be used to locate where cell death occurs in a person's body. (From Kaiser, 2016.)

of circulating MMPs with stage of diseases has already been suggested (see chapters 6,7 and 8), and the idea of linking the levels of MMPs to the cells that are actually producing these MMPs gives another layer of sophistication and prognostication. The additive and/or synergistic effects resulting from combining information on circulating MMPs with that on circulating ccfDNA surely takes the era of precision medicine to new heights.

10.5 Final Thoughts

Five decades after Gross and Lapiere described a "collagenolytic activity" present in the tails of tadpoles undergoing metamorphosis, a large family of related enzymes has emerged. Now called MMPs, members of this family have multiple functions and the importance of these functions in normal human physiology and disease pathologies is widely appreciated. There is still hope that all the information gathered on these enzymes over the past half century, along with new knowledge that continues to accrue, will lead to therapeutic inhibitors that target specific MMPs, with minimal off-target effects. Recent studies on the initial experimental successes with compounds that target exosites suggest that this goal may be within reach. Nonetheless, given the complexity of MMP functions and the diverse mechanisms that regulate MMP expression in normal physiology and disease pathologies, obstacles remain. Even given the availability of an effective therapeutic agent, there are the questions of where and when to use it.

Since we now realize that levels of these enzymes can increase and decrease substantially during different stages of normal physiology and disease pathology, perhaps inhibiting specific MMPs at early stages of disease, before they have irreversibly destroyed the extracellular matrix (ECM) and/or chronically influenced cell behavior as a result of their other functions, may be preferable therapeutically. Developing new therapeutics, coupled with more sensitive techniques to monitor systemic levels of MMPs, may permit the more timely administration of agents that inhibit these enzymes. However, since MMPs have the potential to be beneficial, the risk/benefit ratio needs to be evaluated, an issue that is already well-recognized with many other standard therapeutic regimes.

Another important point is that MMPs represent just one family of proteinases that are upregulated during the course of many diseases and that all of these proteinases may act in concert with MMPs. Conceivably, these enzymes might compensate for MMPs if they were blocked by matrix metalloproteinase inhibitors (MMPIs). Consequently, it may be desirable to use a "protease chip" to determine the "DEGRADOME" of each patient, that is, the complete set of proteases that is produced by a specific diseased tissue at a certain stage of disease progression.[11] Data gathered by these kinds of global analyses may facilitate decisions about the most effective therapeutic approach for blocking proteinases.

Still another challenge is continuing to identify the numerous mechanisms that control MMP synthesis and activity.[11,12] This includes not only positive and negative regulatory transcription factors, but also all of the signal-transduction pathways that function in both stromal and diseased cells. Although complex, the multiplicity of mechanisms that regulate MMPs may present new opportunities for therapeutic intervention, while still considering the temporal–spatial and tissue-specific expression patterns of MMPs.

Developing improved strategies for inhibiting MMPs will require using new techniques that sensitively measure MMPs levels and that assess the efficacy of the different compounds that block MMP production or activity in patients. The new imaging techniques that allow *in situ* measurement of MMPs and their presence in the circulation during the course of therapies could become standard tools for monitoring therapeutic efficacy and disease progression. The underlying issue of understanding when, where and if a particular MMP should, or should not be, inhibited will remain, and will need to be placed in the context of risk/benefit. However, these are challenges are already acknowledged by modern medicine with respect to many medical conditions and applying them to MMPs is a logical step. After all, MMPs are now accepted as crucial contributors to human biology and thus, can rightfully assume their place in the era of precision medicine.[11]

References

1. Brkic, M, S Balusu, C Libert and RE Vandenbroucke (2015). Friends or foes: matrix metalloproteinases and their multifaceted roles in neurodegenerative diseases. *Mediators Inflammation*, 2015, 620581.
2. Hernandez-Guillamon M, S Mawhirt, S Blais, J Montaner, TA Neubert, A Rostagno and J Ghiso (2015). Sequential Amyloid-β Degradation by the Matrix Metalloproteases MMP-2 and MMP-9. *Journal of Biological Chemistry*. 290(24):15078–15091.
3. Moore, CS and SJ Crocker (2012). An alternate perspective on the roles of TIMPs and MMPs in pathology. *American Journal of Pathology*, 180, 12–16.

4. Put, S, R Westhovens, T Lahoutte and P Matthys (2014). Molecular imaging of rheumatoid arthritis: emerging markers, tools, and techniques. *Arthritis Research and Therapy,* 16, 208.
5. Shay, G, CC Lynch and B Fingleton (2015). Moving targets: emerging roles for MMPs in cancer progression and metastasis. *Matrix Biology,* 1–7. doi:10.1016/j.matbio.2015.01.019.
6. Saghatelian, A, N Jessani, A Joseph, M Humphrey and BF Cravatt (2004). Activity-based probes for the proteomic profiling of metalloproteases. *Proceedings of the National Academy of Sciences of the USA,* 101, 10000–10005.
7. Kaiser, BJ (2016). Tests of blood-borne DNA pinpoint tissue damage. *Science,* 351, 1253.
8. Lehmann-Werman R *et al.* (2016). Identification of tissue-specific cell death using methylation patterns of circulating DNA. *Proceedings of the National Academy of Sciences of the USA,* 113, 1826–1834.
9. Sun, K *et al.* (2015). Plasma DNA tissue mapping by genome-wide methylation sequencing for noninvasive prenatal, cancer, and transplantation assessments. *Proceedings of the National Academy of Sciences of the USA.* doi:10.1073/pnas.1508736112.
10. El Messaoudi, S *et al.* (2016) Circulating DNA as a strong multimarker prognostic tool for metastatic colorectal cancer patient management care. *Clinical Cancer Research,* 22, 3067–3077.
11. Eckhard, U, G Marino, GS Butler and CM Overall. (2015). Positional proteomics in the era of the human proteome project on the doorstep of precision medicine. *Biochimie,* 122, 1–9.
12. Overall, CM and C López-Otín (2002). Strategies for MMP inhibition in cancer: innovations for the post-trial era. *Nature Reviews Cancer,* 2, 657–672.

Glossary

3′-untranslated region (3′-UTR): The section of mRNA that immediately follows the codon that stops the translation of the mRNA into a protein. The 3′-UTR contains a poly(A) tail of a string of A's, together with regulatory regions that post-transcriptionally influence gene expression.

5′-untranslated region (5′-UTR): The section of mRNA which is directly upstream from the initiation codon, i.e. the codon that directs the initiation or start of the translation of the mRNA into protein. The 5′ UTR forms a secondary structure that helps to regulate translation.

Acetylation: A process that introduces an acetyl group into histone proteins. The process decreases the positive charge on histones, reducing their binding affinity to DNA, thus facilitating the binding of transcription factors to DNA and enhancing gene expression. **Deacetylation** is the removal of an acetyl group.

Agarose gel electrophoresis: A method used for separating DNA (or RNA) molecules by size. Previously, this method was used to separate RNA molecules by size in preparation for Northern blot analysis of mRNA. Currently, however, mRNA levels are measured by **qRT-PCR** (see definition below).

Allele: Alternative forms of a gene at a given locus.

Allelic heterogeneity: Similar or identical phenotypes caused by different (mutant) alleles at the same genetic locus.

Allostery: The regulation of a protein by binding an effector molecule at a site other than the enzyme's active site.

Antigen: Any molecule that provokes the synthesis of an antibody (humoral immunity) and/or the activation of T cells (cellular immunity).

Antisense strand (of DNA): The non-coding strand of double-stranded DNA. It is complementary to the mRNA and serves as the template for RNA synthesis.

Base pair: In double stranded DNA there is complementary hydrogen bonding (adenine–thymine and guanine–cytosine). The hydrogen-bonded residue pair is designated "1 bp". This unit is used for measuring the length of pieces of DNA.

Chromosome: The location of hereditary (genetic) material within a cell. This hereditary material is packaged in the form of a very long double-stranded molecule of DNA surrounded by, and complexed with, several different forms of protein.

Cis element: A response element in DNA that binds transcription factors to activate or suppress gene expression.

Cloning: The generation of a large number of identical DNA components or cells from a single copy.

Codon: A group of three sequential nucleotides within RNA that codes for a specific amino acid in the corresponding protein. Codons are "words" of the genetic code (e.g. AUG = methionine, and also means "start translation here").

Complementarity: The specific binding of adenine to thymine (or uracil in RNA) and cytosine to guanine on opposite strands for DNA or RNA.

Complementary DNA (cDNA): DNA synthesized from an mRNA template. cDNA is often used as a probe to help locate a specific gene in DNA and/or to bind to corresponding mRNA to measure mRNA levels in cells.

Deacetylation: See **acetylation.**

Denature: To separate two strands of nucleic acids by heating them to a temperature sufficient to break the G/C, A/T or A/U bonds. Denaturing does not destroy individual strands and they can re-nature as the temperature is lowered. The process of re-naturing is the basis of nucleic acid hybridization. Denaturing proteins, on the other hand, will destroy secondary/tertiary structure, which prevents refolding/renaturing.

Disintegrins: A family of small proteins (45–84 amino acids in length) from snake venom, which are potent inhibitors of both platelet aggregation and integrin-dependent cell adhesion.

DNA (deoxyribonucleic acid): The molecule containing hereditary information in most organisms (some viruses use RNA). The molecule is double-stranded, with an external backbone formed by a chain of alternating phosphate and sugar (deoxyribose) units and an internal ladder-like structure formed by nucleotide base-pairs held together by hydrogen bonds. The nucleotide base-pairs consist of bases adenine (A), cytosine (C), guanine (G) and thymine (T), whose structures are such that A can bond only with T, and C only with G. The sequence of each individual strand can be deduced by knowing that of its partner. The complementarity is the key to information-transmitting capabilities of DNA, and its ability to be replicated.

DNA polymerase: The enzyme that synthesizes DNA from a complementary DNA template. In the cell, it replicates the DNA strand prior to cell division.

DNA probe: A nucleic acid molecule of known structure and/or function that has been tagged with a tracer substance (radioactive isotope or fluorescent dye) that is used to locate and identify a specific gene or region of a chromosome or portion of the genome. It may also be used to identify a specific mRNA. DNA probes are used for detection on Southern (named for Dr. E. Southern; DNA) or Northern (rarely still used; RNA) blots.

Domain: A region of amino acid sequence of a protein that can be equated with a particular function, or a corresponding segment of a gene. For example, Matrix Metalloproteinases (MMPs) have a "catalytic domain" that mediates their proteolytic activities.

Endonuclease: An enzyme that cleaves specific sequences within a DNA or RNA strand, or within specific amino acid sequences of a protein.

Exon: Sequences within the DNA that are represented (expressed) in the final mature RNA.

Exosite: A secondary binding site, remote from the active site, on an enzyme. This is similar to an allosteric site, but differs in the fact that, in order for an enzyme to be active, its exosite typically must be occupied.

Gene: The portion of a DNA molecule that is sufficient for the expression of a functional polypeptide (protein).

Gene amplification: A process that increases the number of copies of the same gene within a cell. Amplification may be spontaneous or induced.

Genetic engineering: A technique used to modify the genetic information in a living cell, reprogramming it for a desired purpose, such as the production of a protein it would not naturally produce.

Genetic locus: A specific position or location on a chromosome.

Genome: The complete genome of an organism, containing its complete genetic information. The human genome consists of about 100,000 genes, and about 1 billion base pairs (bp) of DNA.

Heterozygote (heterozygous): An individual who has two different alleles at a given local on a pair of homologous chromosome.

Homology: Similarity between two distinct genes in their nucleotide sequences.

Homozygote (homozygous): An individual who has two identical alleles at a given locus on a pair of homologous chromosomes.

Hybridization: Pairing of an RNA and a DNA strand or of two different DNA strands. Efficient hybridization requires a high degree of complementarity of the two strands.

Immunoblot: See **western blot.**

Intron: Sequences within the DNA of a gene that are not represented in the final RNA. They are removed from the precursor RNA within the nucleus, and the remaining pieces of the RNA are spliced together prior to export out of the nucleus.

***In situ* hybridization:** The localization of RNA sequences by hybridization of labeled probes to tissue sections. It is used to identify the cell type that expressed a given gene, e.g. MMP-1 mRNA in fibroblasts.

Interleukin-4: A cytokine that induces differentiation of naive helper T cells (Th0 cells) to Th2 cells. The Th2 cells produce additional IL-4 in a positive feedback loop.

Interferons: A group of signaling proteins made and released by host cells in response to the presence of several pathogens, such as viruses and bacteria. **Interferon gamma (IFNγ)** is critical for immune responses to viral, some bacterial and protozoal infections. It is an important activator of macrophages.

Kilobase (kb): One thousand base pairs in a DNA sequence.

Ligase: An enzyme that joins fragments of DNA. It is used to form recombinant molecules.

Messenger RNA (mRNA): An RNA transcribed from the DNA of a gene, forming the template from which a protein is translated.

Methylation: The process in which methyl groups are added to cytosine in DNA. When methylation of cytosine occurs on or near the promoter region of a gene, expression of the gene is suppressed.

Micro-RNAs (mi-RNAs): A group of non-coding RNAs that have key roles in the regulation of gene expression. They bind to specific sequences in specific mRNAs and target these mRNAs for degradation or translational repression.

Missense mutation: A single DNA base substitution resulting in a codon specifying a different amino acid.

Monoclonal antibodies: Antibodies produced by a single source (clone) of cells that only recognize a single antigen.

Nuclear localization sequence/signal (NLS): An amino acid sequence that "tags" a protein for import into the cell nucleus by nuclear transport.

Nucleophilic attack: A chemical reaction where a nucleophile (a reactant) provides a pair of electrons to attack a covalent bond to form a new covalent bond with an amino acid.

Pallindromic sequence: A nucleotide sequence that reads the same in either direction ("Madam I'm Adam"). Restriction endonuclease recognition sites in DNA are palindromes.

Phenotype: The appearance or characteristics of an organism that results from the interaction between its genotype and the environment.

Point mutation: Substitution of one nucleotide for another in DNA.

Polymerase chain reaction (PCR): Enzymatic technique for the production of many copies of one DNA sequence in a test tube. Repeated cycles of temperature changes allow sequential denaturation, priming, and strand synthesis steps that result in an exponential production of the desired DNA molecule.

Polymorphism: In a population, the common occurrence of two or more genetically determined alternative phenotypes.

Pre-pro-protein: Protein precursors that are secreted from the cell are synthesized with an N-terminal "signal peptide" that targets them for secretion, and these are called pre-pro-proteins. The signal peptide is cleaved off inside the cell (in the endoplasmic reticulum) before the protein is secreted.

Pro-protein: Protein precursors that are in an inactive form but can become an active protein by post-translational modification, such as cleaving off a piece of the protein molecule. Enzyme precursors are called zymogens or proenzymes.

Probe: A labeled DNA or RNA sequence used to detect the presence of a complementary sequence by molecular hybridization.

Promoter: Sequence of DNA giving the signal to start (or promote) transcription. The promoter is found upstream (i.e. 5′) of the gene, and is important in the regulation of gene expression.

Receptor: Protein that binds specific hormones or other types of ligands and initiates physiological functions.

Response element: A short specific sequence of DNA, usually located in the promoter region of a gene, to which a specific transcription factor(s) binds and influences the rate of gene expression.

Reverse transcriptase: An enzyme that synthesizes DNA from an RNA template.

Reverse transcription polymerase chain reaction (RT-PCR): A variation of the polymerase chain reaction (PCR). The technique of RT-PCR is used to detect mRNA expression.

qRT-PCR: A version of RT-PCR that quantitatively measures the amplification of DNA using fluorescent probes. The technique is used to quantify mRNA. Copies of mRNA are first amplified by reverse transcriptase to "convert" mRNA transcripts to DNA, which is then amplified by PCR.

Ribonucleic Acid (RNA): A polynucleotide consisting of a backbone of alternative phosphate and sugar (ribose) molecules to which are attached the nucleotide bases adenine (A), thymine (T), guanine (G) or uracil (U), which replaces the cytosine (C) of DNA. There are several classes of RNA that have different purposes: messenger RNA (mRNA); micro-RNA (mi-RNA); transfer RNA (tRNA); ribosome rRNA.

RNA polymerase: The enzyme that synthesizes RNA from a DNA template.

Signal transduction: A process in which an extracellular signaling molecule activates a specific receptor located on the cell surface or inside the cell. This receptor triggers a biochemical chain of events (usually a series of phosphorylation events) inside the cell that results in a change in cellular biochemistry/behavior.

TATA box: A specific sequence of nucleotides located immediately upstream from the transcriptional start site, it serves to position the RNA polymerase II complex to affect transcription.

Template: The nucleotide sequence that is "read" by DNA or RNA polymerases in order to synthesize a complementary nucleic acid strand.

Th1 helper cells: Th1 cells primarily produce interferon gamma (IFNγ) and interleukin (IL)-2, while Th2 cells produce IL-4, IL-5,

IL-6, IL-10, and IL-13. The two helper T cell classes also differ by the type of immune response they produce. Th1 cytokines stimulate macrophages, lymphocytes, and PMNs in the destruction of bacterial pathogens. These cytokines also help foster the development of cytotoxic lymphocytes (cytotoxic T lymphocyte [CTL] and natural killer [NK] cells) that are responsible for the cell-mediated immune response against viruses and tumor cells.

Th2 helper cells: Th2 cells mediate the activation and maintenance of the humoral (antibody-mediated) immune response against extracellular parasites, bacteria, allergens, and toxins. Th2 cells facilitate these functions by producing cytokines (e.g. IL-4, IL-5, IL-6, IL-9, IL-13) that mediate antibody production and eosinophil activation.

Thrombospondins (TSP): Secreted proteins with anti-angiogenic abilities.

Transcription: The synthesis of RNA on a DNA template. The information within the DNA is transcribed into multiple RNA copies.

Transfection: The process of forcing cells to take up DNA from the external environment.

Transgenic: A state in which an organism contains, within its germline genome, both parental and foreign DNA sequences. The foreign sequences are thus transmissible to the offspring.

Translation: The synthesis of a polypeptide with an amino acid sequence specified by the codon sequences of a corresponding mRNA.

Upstream and downstream: Terms that refer to a relative position in DNA or RNA. Each strand of DNA or RNA has a 5′ end and a 3′ end, so named for the carbon position on the deoxyribose (or ribose) ring. Upstream is toward the 5′ end of the RNA or DNA molecule and downstream is toward the 3′ end.

Western blot (also known as immunoblot): A blotting technique analogous to Southern blotting which detects DNA, for detecting proteins, usually detected with antibodies as "probes."

Wnt signal transduction pathways: These pathways were originally described in *Drosophila*, and their names reflect their origins from these fruit flies. Signal transduction through Wnt occurs through cell surface receptors: (1) the canonical (i.e. the first described); (2) the non-canonical pathway; and (3) the noncanonical Wnt/calcium pathway. All are activated by binding of a Wnt-protein ligand to a receptor (called Frizzled), which sends the signal to the "dishevelled" protein inside the cell. The canonical Wnt pathway helps regulate gene transcription, while the non-canonical planar pathway regulates the cytoskeleton and the shape of the cell. The non-canonical Wnt/calcium pathway regulates calcium inside the cell. Wnt signaling pathways are highly conserved from fruit flies to humans.

Zymography: An electrophoretic technique for detecting the activity of enzymes. Samples are prepared in a standard, non-reducing loading buffer and proteins are separated by polyacrylamide gel electrophoresis (PAGE). No reducing agent or boiling is necessary since these would interfere with refolding of the enzyme and retention of proteolytic activity. A suitable substrate such as gelatin or casein for protease detection is embedded during preparation of the acrylamide gel. Following electrophoresis, the gel is incubated for an optimized length of time at 37°C. The zymogram is then stained and areas of digestion appear as clear bands against a darkly stained background where the substrate has been degraded by the enzyme.

Index

$\alpha 2\beta 1$ integrin, 114
acetylation, 32, 56
activating protein, 19
activator protein-1 (AP-1), 33
activity-based probes (ABP), 259
ADAMs, 230
ADAMTS, 230
allosteric, 239
allostery, 239
alpha2 macroglobulin (α2M), 10
Alzheimer's amyloid-β (Aβ), 255
aminophenylmercuric acid (APMA), 27
aneurysm, 188, 195–197, 203–206
aneurysmal, 189
angiogenesis, 160
AP-1 site, 242
apoptosis, 160
arthritis, 227
aspartic, 1
atherosclerosis, 188–191, 193, 194, 196, 203–205
autoimmune/inflammatory, 245

basement membrane, 17
biologic therapies, 245
biomarkers, 166
bone, 17, 97

cancer, 33, 77, 227
cardiovascular disease, 33, 227
cartilage, 17
catalytic, 25
catalytic domain, 239
catalytic site, 232
cDNAs, 25
central nervous system (CNS), 255
chemokines, 117
chondrocytes, 77, 97
chromosome, 20
chronic venous disease (CVD), 117
chronic wounds, 33
circulating cell-free DNA (ccfDNA or ccDNA), 262
clones, 25
coding region, 31
collagen, 114, 124, 125, 128, 129, 133, 138, 140, 142
collagenase, 33, 96, 124–128, 131
collagen types I and III, 13
coronary artery disease, 203, 207
crystal structure, 25
C-terminal hemopexin-like, 26

cysteine, 1
cysteine switch, 26
cytokine, 38, 65, 96, 117, 124, 127, 129, 132, 242, 254

DEGRADOME, 264
disintegrin and metalloproteinase (ADAMS), 6
disintegrin and metalloproteinase with thrombospondin motif (ADAMTS), 6
DNA, 31
domain, 25, 26

early response genes, 44
endogenous inhibitors, 31
endopeptidases, 2
endothelial cells (ECs), 77
enzymatic activities, 31, 254
epigenetic mechanisms, 32
epithelial-to-mesenchymal transition (EMT), 159
erythroblastosis twenty six site (ETS), 35
exons, 20, 31
exosites, 27, 232
exosomes, 75
extracellular matrix (ECM), 65, 98, 253

fibroblastic growth factor (FGF), 160
fibrosis, 124, 136–141

gene expression, 231, 254
gene regulation, 96
genomic, 25
genomic clones, 33
glucocorticoids, 44, 242
G protein-coupled receptors (GPCRs), 85
growth factors, 38, 65, 96, 117, 242
GS-5745, 173

hear failure (HF), 188, 194, 201–205
hemopexin domain, 239
heterogeneous nuclear ribonucleic acid (hnRNA), 31
histones, 32
homeostasis, 95, 230
hormones, 85
hypoxia, 161

immunity, 107
  immune response, 103, 110, 113, 114, 116
  immune system, 107
inflammation, 113, 159
inflammatory/autoimmune, 254
integrin α2β1, 73
interleukin-1 (IL-1), 124, 126–131, 254
interleukin-1 receptor antagonist (IL-1Rα), 131–132
interleukin-1β (IL-1β), 35, 96
intracellular, 65
Intracellular Localization, 79
introns, 20, 31

keratinocyte, 72, 114

linker domain, 26
lysosomes, 2

matrix metalloproteinase inhibitors (MMPIs), 95, 229, 264
matrix metalloproteinases (MMPs), ix, 1, 31, 65, 95, 227, 253
metabolic syndrome, 88
metalloproteinases, 2
methylation, 32, 58
microenvironment, 66, 107
micro-RNAs (miRNAs), 32
mitogen activated protein kinases (MAPKs), 41
MMP-1, 124, 125, 127–130, 133–140, 142
mmp1α, 37, 103
MMP knockout mice, 103
MMP–tissue inhibitors of metalloproteinase (TIMP), 257
molecular imaging probes, 258
monoclonal antibody, 239
monocyte-macrophages, 128
mononuclear cell factor (MCF), 126, 127
mRNA, 31
mRNA stability, 51
murine MMP-1, 103
myocardial infarction, 188, 197, 198, 204–206

neoangiogenesis, 116
nuclear factor-κB (NF-κB), 45, 159
nuclear localization, 82
nucleophilic attack, 2

osteoarthritis (OA), 33, 124, 129, 130, 133, 134, 136, 142, 236

pericellular, 65, 68
Periostat® (doxycycline hyclate), 172
polymorphisms, 162
polyomavirus enhancer activator 3 [PEA3], 35
posttranscriptional modification, 31
procollagenase, 19
promoter, 31
propeptide, 25
prostaglandin E2 (PGE2), 125–129, 131, 137
protease-activated receptors (PARs), 90
protease chip, 264
protease web, 66
proteinases, 1
proteoglycanase, 19

RARs α, β, and γ, 49
receptors, 65
retinoic acid receptors, 49
retinoids, 44, 242
retinoid X receptors (RXRs α, β, and γ), 49
rheumatoid, 33, 236
rheumatoid arthritis (RA), 17, 96, 124, 125, 127, 131–136, 140–143, 254

serine, 1
signaling molecules, 85
signal peptide, 25
signal transduction, 38, 85, 96, 242, 253
single-nucleotide polymorphisms (SNPs), 46

skeletal and cardiac muscle, 97
small molecule chemical inhibitors, 227
stromelysin, 19
subsites, 231
synovial cells, 124, 127, 128, 131
synovium, 96

TATA box, 37
tetracycline, 238
threonine, 1
thrombin, 90
tissue inhibitors of metalloproteinases (TIMPs), 9
transcription, 31
transcription factor, 31, 84, 253
transforming growth factor-beta (TGF-β), 43, 160
transin-1, 19
triple helix, 13
tumorigenesis, 158
tumor imaging, 170
tumor microenvironment, 168
tumor necrosis factor (TNF), 124, 127–132, 135, 142
tumor suppressor, 110
tumour necrosis factor-alpha (TNF-α), 254
type I collagen, 73

3′-untranslated region (3′-UTR), 31

vascular endothelial growth factor (VEGF), 160

wound, 113
wound healing, 72

zymogen, 18